The Illusion Of Reality

How Humanity Constructs Its Own Universe

By: Joshua Douthett

Prologue: The Fabric of Illusion

Understanding Reality

Reality, as we perceive it, is a complex tapestry woven from the threads of human cognition, culture, language, and shared beliefs. This intricate web shapes our understanding of the world, dictating how we interpret experiences, interact with others, and form our identities. Yet, what if everything we know and hold as real is merely a construct—a grand illusion crafted by our minds and societies? This idea, though unsettling, invites us to question the very foundation of our existence and the structures we rely on to make sense of it. It challenges us to consider that our perceptions, deeply influenced by cultural norms and linguistic frameworks, might not reflect an objective reality but rather a subjective construct shaped by collective human consciousness.

This book delves into the depths of this provocative idea, exploring how every aspect of our perceived reality is a creation of humanity. From the languages we speak to the mathematical principles we use to understand the universe, each element of our knowledge is a product of human invention. Language, for instance, is not merely a tool for communication but a lens that shapes our thought processes and worldviews. Different languages encode different realities, influencing how their speakers perceive time, space, and relationships. Similarly, mathematics, often regarded as the purest form of objective truth, is rooted in human-defined axioms and conventions. Its beauty and consistency lie not in an inherent universal order but in the logical structures we have developed to describe patterns and relationships.

The idea that reality is a construct extends to the social and cultural dimensions of our lives. Social norms, values, and institutions are all products of collective human agreements, constantly evolving and reshaping our interactions and identities. Money, for example, is a powerful social construct that holds no intrinsic value yet governs economies and personal livelihoods due to shared belief in its worth. Likewise, concepts of morality and ethics, which guide our sense of right and wrong, are deeply influenced by cultural contexts and historical developments. These constructs, though intangible, exert a profound influence over our behaviors and societal structures, reinforcing the notion that much of what we consider real is built upon collective human imagination.

Furthermore, scientific paradigms, which we often consider the pinnacle of objective understanding, are also subject to human constructs. The scientific method itself is a framework devised to investigate and interpret phenomena, relying on empirical evidence and reproducibility. However, the history of science reveals that our understanding of the natural world is provisional, subject to change with new discoveries and shifting paradigms. This highlights the constructed nature of scientific knowledge, as theories and models are tools to navigate and explain the complexities of the universe, not definitive representations of an ultimate reality.

By examining these constructs, we gain insight into the profound influence of human thought on our understanding of existence. This exploration does not aim to undermine the value of human knowledge and achievements but to provide a perspective that recognizes the fluidity and subjectivity of reality. It encourages us to approach our beliefs and knowledge systems with a critical and reflective mindset, appreciating the creativity and ingenuity behind their construction. In doing so, we can foster a deeper understanding of ourselves and the world we inhabit, navigating the complexities of existence with greater awareness and flexibility.

The Purpose of This Book

The aim of this book is not to undermine the achievements of human knowledge but to provide a perspective that challenges the conventional understanding of reality. Human accomplishments in various fields—from the arts and humanities to science and technology—are testaments to our remarkable capacity for innovation and discovery. These achievements have enriched our lives, expanded our horizons, and driven progress. However, by taking a step back and critically examining the foundations upon which these accomplishments are built, we can gain a more nuanced understanding of what we consider to be real.

By examining the constructs that form our world, we can gain a deeper appreciation of the power of human thought and the fluid nature of reality. This perspective allows us to see that our systems of knowledge are not static or absolute, but dynamic and evolving. Each new discovery, each paradigm shift, reflects the ongoing process of constructing and reconstructing our understanding of the world. This fluidity is a source of strength, enabling us to adapt to new challenges and incorporate new insights into our collective knowledge.

Moreover, this book aims to inspire curiosity and open-mindedness. In recognizing the constructed nature of our reality, we can become more receptive to alternative viewpoints and more willing to question established norms and beliefs. This openness is crucial in a world that is increasingly interconnected and complex, where rigid adherence to outdated constructs can hinder progress and understanding. By embracing the fluidity of reality, we can foster a more inclusive and adaptable approach to knowledge.

This exploration also underscores the importance of creativity and imagination in shaping our world. The constructs we rely on—whether linguistic, mathematical, social, or scientific—are products of human creativity. They reflect our ability to envision new possibilities and find innovative solutions to problems. By appreciating the role of imagination in constructing reality, we can cultivate a more creative and forward-thinking mindset, one that values innovation and embraces change.

Finally, this book encourages a deeper awareness of the ethical implications of our constructs. Understanding that our realities are shaped by collective agreements highlights the responsibility we have in shaping these constructs. Social norms, economic systems, and scientific paradigms all carry ethical dimensions that affect individuals and societies. By critically examining these constructs, we can work towards creating more just and equitable systems that reflect our highest values and aspirations.

Chapter 1: The Concept of Reality

Defining Reality

Reality is often defined as the state of things as they actually exist, as opposed to an idealistic or notional idea of them. However, this definition is itself a construct, subject to interpretation and understanding by human minds. Various philosophies offer differing views on what constitutes reality, from materialism, which posits that only physical matter is real, to idealism, which suggests that reality is fundamentally mental or spiritual in nature. These differing viewpoints highlight the complexities and subjective nature of understanding reality.

Materialism asserts that reality consists solely of matter and energy, and that all phenomena, including consciousness and thought, can be explained in terms of physical processes. This perspective is grounded in the scientific method and empirical evidence, relying on observable and measurable phenomena to define what is real. From the materialist viewpoint, the physical universe is the ultimate reality, and everything else, including human experiences and emotions, is a result of interactions between physical entities.

On the other hand, idealism proposes that reality is primarily constructed by the mind. According to this view, the material world is secondary to the mental or spiritual realm. Philosophers like George Berkeley argued that objects only exist insofar as they are perceived by a conscious observer. In this sense, reality is a construct of our perceptions and consciousness. Idealism emphasizes the role of the mind in shaping our experiences and posits that without perception, there can be no reality.

Between these two extremes lie various other philosophical perspectives that blend elements of both materialism and idealism. For instance, dualism, as proposed by René Descartes, posits the existence of both physical and non-physical substances, suggesting that reality comprises both material and mental elements. Meanwhile, phenomenology, championed by Edmund Husserl, focuses on the structures of consciousness and the phenomena that appear in acts of consciousness, arguing that reality is accessible through subjective experience.

Philosophical Perspectives

Throughout history, several philosophers have grappled with the concept of reality, each contributing to our understanding of this elusive notion. Plato's Allegory of the Cave is one of the earliest and most influential explorations of reality and illusion. In this allegory, prisoners in a cave perceive shadows on a wall as the entirety of reality, unaware of the true forms that cast these shadows outside the cave. Plato uses this story to illustrate the difference between the world of appearances, which is deceptive and limited, and the world of forms, which represents true and unchanging reality.

René Descartes, a key figure in modern philosophy, approached the question of reality through methodological skepticism. In his Meditations, Descartes famously doubted everything that could be doubted, ultimately arriving at the conclusion, "Cogito, ergo sum" ("I think, therefore I am"). This statement underscores the idea that the act of thinking is undeniable proof of one's own existence, highlighting the role of the mind in constructing reality.

Immanuel Kant offered another profound perspective on reality. He distinguished between the noumenal world (the world as it is in itself) and the phenomenal world (the world as we experience it). According to Kant, our knowledge is limited to the phenomenal world, as our senses and cognitive faculties shape our experiences. The noumenal world, on the other hand, remains beyond our direct comprehension. This distinction underscores the idea that reality, as we know it, is shaped by the structures of human perception and cognition.

The Role of Perception

Perception plays a crucial role in shaping our understanding of reality. Our brains construct a version of reality based on sensory inputs, but this version is inherently subjective. Factors such as cognitive biases, cultural background, and personal experiences influence how we perceive the world, leading to the conclusion that our reality is, to a significant extent, a mental construct.

One striking example of the subjectivity of perception is the phenomenon of optical illusions. These visual tricks occur when our brains misinterpret sensory information, leading us to perceive something that does not correspond to objective reality. Optical illusions demonstrate that our senses can be deceived and that what we perceive is not always an accurate representation of the external world.

Cognitive biases also play a significant role in shaping our perception of reality. These biases are systematic patterns of deviation from norm or rationality in judgment, often influenced by our experiences, emotions, and social context. For instance, confirmation bias leads us to favor information that confirms our preexisting beliefs, while ignoring or dismissing evidence that contradicts them. This bias can create a skewed perception of reality, reinforcing our beliefs and misconceptions.

Moreover, cultural background profoundly impacts how we perceive and interpret reality. Different cultures have unique ways of understanding and interacting with the world, shaped by their languages, traditions, and social norms. For example, the concept of time varies significantly across cultures, with some viewing it as a linear progression, while others see it as cyclical or fluid. These cultural differences highlight the diversity of human experiences and the constructed nature of our perception of reality.

Personal experiences also shape our perception of reality. Our memories, emotions, and past interactions influence how we interpret new information and experiences. This subjective lens means that two individuals can perceive the same event in vastly different ways, each constructing their own version of reality based on their unique perspectives.

In conclusion, the concept of reality is a complex and multifaceted construct, shaped by philosophical perspectives, perception, and subjective experiences. By exploring the various ways in which reality is defined and understood, we can appreciate the fluid and constructed nature of our perceptions, gaining deeper insight into the nature of existence and the power of human thought in shaping our world.

Chapter 2: Language - The First Construct

The Birth of Language

Language is one of humanity's most profound creations. It allows us to communicate complex ideas, share knowledge, and build societies. However, language also shapes our reality by providing the framework through which we interpret and understand the world. The development of language marked a significant milestone in human evolution, enabling us to convey abstract concepts, emotions, and experiences. This powerful tool has not only facilitated social cohesion but has also influenced how we perceive and construct reality.

The origins of language are deeply rooted in our need for social interaction and cooperation. Early humans likely developed rudimentary forms of communication through gestures, sounds, and facial expressions. Over time, these forms of communication evolved into more sophisticated systems of spoken language. As societies grew more complex, so did their languages, eventually leading to the development of written language. This progression allowed for the preservation and transmission of knowledge across generations, further solidifying the role of language in shaping human reality.

Language as a Tool and a Limitation

While language enables communication, it also imposes limitations. Words can only approximate reality, often failing to capture the full essence of experiences or phenomena. This linguistic relativity implies that our understanding of reality is constrained by the language we use. For example, the word "love" encompasses a wide range of emotions and experiences, yet it falls short of fully expressing

the depth and complexity of the feeling.
Similarly, abstract concepts like "time" or
"beauty" can be difficult to articulate precisely,
as their meanings can vary significantly
depending on cultural and individual
interpretations.

Furthermore, the structure of a language can
influence how its speakers perceive the world.
For instance, some languages have multiple
words for different types of snow, reflecting the
cultural significance of snow in those societies.
In contrast, languages spoken in tropical regions
may lack such distinctions. This phenomenon,
known as linguistic determinism, suggests that
the language we speak can shape our thought
processes and perceptions of reality.

The Sapir-Whorf Hypothesis

The Sapir-Whorf Hypothesis, also known as linguistic relativity, posits that the structure of a language affects its speakers' worldview and cognition. This hypothesis, named after linguists Edward Sapir and Benjamin Lee Whorf, suggests that people who speak different languages perceive and think about the world differently. According to this view, language is not just a tool for communication but a lens through which we experience and interpret reality.

One classic example of linguistic relativity is the way different languages describe spatial relationships. In some languages, such as English, speakers use relative terms like "left" and "right" to describe spatial orientation. In contrast, speakers of languages like Guugu Yimithirr, an Australian Aboriginal language, use cardinal directions (north, south, east, west) regardless of their orientation. This fundamental difference in linguistic structure influences how speakers of each language navigate and perceive their environment.

Another example is the concept of time. In English, time is often conceptualized as a linear progression, with a clear past, present, and future. However, in some languages, such as Aymara, spoken in the Andes, the past is perceived as being in front of the speaker, while the future is behind. This linguistic difference reflects a unique cultural understanding of time and highlights the influence of language on cognitive processes.

How Language Shapes Thought

Language influences thought processes, categorization, and memory. For instance, cultures with numerous words for snow perceive it differently than those with only one. This linguistic influence extends to abstract concepts like time, space, and emotion, demonstrating that language not only describes but also shapes our reality.

Research in cognitive science supports the idea that language shapes thought. Studies have shown that bilingual individuals often think differently depending on the language they are using. For example, bilinguals who speak languages with different color terminologies may categorize colors differently based on the language context. This phenomenon suggests that language can influence perceptual processes and cognitive categorization.

Memory is another area where language plays a crucial role. The way we encode and recall experiences is often influenced by the language we use to describe them. For instance, the use of specific vocabulary and grammatical structures can affect how we remember events and construct narratives about our lives. This influence of language on memory underscores its role in shaping our personal and collective histories.

Language and Reality Construction

Language not only reflects reality but also actively constructs it. Through the use of metaphors, narratives, and discourse, language shapes our understanding of the world and our place within it. Metaphors, for instance, allow us to conceptualize abstract ideas in concrete terms, influencing how we think and act. The metaphor of "time as money," prevalent in Western cultures, shapes how people perceive and manage their time, treating it as a valuable resource that can be spent, saved, or wasted.

Narratives and stories are another way language constructs reality. The stories we tell ourselves and others shape our identities, beliefs, and values. Cultural narratives, such as myths, legends, and historical accounts, play a significant role in forming collective identities and shaping societal norms. These narratives provide a framework for understanding the world, guiding behavior, and reinforcing cultural values.

Discourse, the way language is used in social contexts, also plays a crucial role in constructing reality. Through discourse, power dynamics are established, maintained, and challenged. The language used in political, legal, and educational settings, for example, shapes public opinion, policy decisions, and societal structures. Critical discourse analysis examines how language reflects and perpetuates power relations, highlighting the role of language in constructing social realities.

Language is a powerful tool that enables communication, shapes thought, and constructs reality. It allows us to share knowledge, express emotions, and build societies. However, language also imposes limitations, as words can only approximate reality and are constrained by cultural and cognitive biases. The Sapir-Whorf Hypothesis and research in cognitive science support the idea that language influences thought and perception, highlighting the complex relationship between language and reality.

By recognizing the constructed nature of language, we can develop a more critical and reflective approach to understanding the world. We can appreciate the diversity of human experiences and the ways in which different languages and cultures shape perceptions of reality. This awareness can foster greater empathy and open-mindedness, encouraging us to question established norms and beliefs and embrace alternative viewpoints.

In exploring the power and limits of language, we gain insight into the fluid and dynamic nature of reality. We come to understand that our perceptions and knowledge are not fixed but are continually shaped and reshaped by the languages we speak and the cultures we inhabit. This perspective invites us to engage with the world with greater curiosity and creativity, recognizing the profound influence of language in constructing the realities we navigate every day.

Chapter 3: Mathematics - The Universal Illusion

The Origins of Mathematics

Mathematics is often seen as the universal language, a pure and objective means of understanding the world. However, its origins lie in human efforts to quantify and make sense of their surroundings, making it a construct born out of necessity and utility.

The origins of mathematics can be traced back to ancient civilizations, where early humans began to use rudimentary forms of counting and measurement as part of their daily activities. The earliest evidence of mathematical activity is found in the form of notched bones and tally sticks, dating back to the Upper Paleolithic period, around 30,000 BC. These artifacts indicate that prehistoric humans used simple arithmetic to track quantities, such as the number of days or objects.

As human societies evolved, so did their mathematical knowledge. Ancient Mesopotamia and Egypt are often credited with significant early contributions to the field. In Mesopotamia, around 3000 BC, the Sumerians developed a system of writing known as cuneiform, which included numerical notation for administrative and trade purposes. They used a sexagesimal (base-60) system, which has influenced our current measurement of time and angles. The Egyptians, around the same period, utilized mathematics for architectural and engineering projects, most notably in the construction of the pyramids. Their mathematical prowess included an understanding of basic geometry and arithmetic operations.

The development of mathematics took a major leap forward in ancient Greece, where scholars began to explore the subject in a more theoretical and abstract manner. Figures such as Pythagoras, Euclid, and Archimedes laid the foundations for various branches of mathematics, including geometry, number theory, and mathematical physics. Euclid's work "Elements" became one of the most influential mathematical texts, systematically presenting the principles of geometry. Greek mathematics was characterized by a rigorous approach to proofs and logical reasoning, setting the stage for future advancements.

Mathematics continued to flourish through the Middle Ages and the Renaissance, with significant contributions from Islamic scholars during the Islamic Golden Age. Mathematicians such as Al-Khwarizmi, whose works introduced algebra, and Omar Khayyam, who made advancements in geometry, played crucial roles in the development of mathematical thought.

31

The transmission of this knowledge to Europe via translations of Arabic texts spurred further progress, leading to the mathematical renaissance in the 16th and 17th centuries. This period saw the emergence of great mathematicians like Newton and Leibniz, who developed calculus, profoundly impacting science and engineering.

The origins of mathematics are deeply rooted in the practical needs of early human societies, evolving through significant contributions from various ancient civilizations and reaching new heights with the formalization of mathematical concepts in ancient Greece. The rich legacy of mathematical development was further enriched by Islamic scholars and Renaissance thinkers, ultimately shaping the sophisticated and diverse field of mathematics we know today.

Mathematics as a Construct

Mathematics, often heralded as a universal language, is a remarkable human construct designed to make sense of the world around us. Far from being inherent to the universe itself, mathematics is a product of human intellect—a system we devised to quantify, measure, and understand the patterns and relationships in nature. Its effectiveness lies in its ability to provide precise descriptions and predictions, but its foundation rests solely on human invention. While it provides a powerful framework for understanding patterns and relationships, its foundations are based on human-devised axioms and logical structures, making it, in essence, a human-made construct.

The concept that mathematics is not a universal language challenges the notion that it exists independently of human cognition. Rather, it reflects our attempt to impose order and structure upon the chaotic complexity of reality. Numbers, equations, and geometric shapes hold no inherent meaning outside of our interpretation and application of them. They are tools we use to navigate and manipulate our surroundings, facilitating everything from engineering marvels to abstract theoretical frameworks.

Moreover, the idea that math "doesn't mean anything" highlights its abstract nature. Unlike natural languages that directly convey thoughts and emotions, mathematics operates in a realm of pure abstraction, divorced from subjective experience. Its symbols and rules have no intrinsic significance beyond what we assign to them. This inherent arbitrariness underscores the human-centric nature of mathematics—it serves our needs for understanding and control, but it does not inherently embody universal truths or principles.

In essence, while mathematics serves as a powerful tool for modeling and understanding the universe, it remains a construct of the human mind rather than a universal language spoken by nature itself. Its utility and elegance lie in its versatility and precision within the realm of human experience, but its meaning and significance derive entirely from our collective agreement and interpretation.

Gödel's Incompleteness Theorems

Gödel's Incompleteness Theorems, formulated by the Austrian mathematician Kurt Gödel in the 1930s, are fundamental results in mathematical logic that have profound implications for the foundations of mathematics. There are two main theorems:

1. **First Incompleteness Theorem:** This theorem states that in any formal system

of mathematics that is sufficiently strong to express arithmetic (such as Peano arithmetic), there exist statements that are true but cannot be proven within the system itself. In other words, there are mathematical truths that lie beyond the reach of any formal proof based on the system's axioms.

2. **Second Incompleteness Theorem:** Building upon the first theorem, the second theorem asserts that if a formal system is consistent (meaning it does not lead to contradictions), then the system cannot prove its own consistency. In simpler terms, if a system is complex enough to express arithmetic and is consistent, it cannot prove that it will never produce a contradiction.

These theorems revolutionized the field of mathematical logic by demonstrating the inherent limitations and boundaries of formal systems. They challenged the optimistic belief prevalent in the early 20th century that mathematics could be entirely formalized and reduced to a finite set of axioms and rules. Gödel's work showed that there are inherent gaps in any sufficiently powerful mathematical system and that some truths are inaccessible to formal proofs within those systems.

The implications of Gödel's Incompleteness Theorems extend beyond mathematics into philosophy, computer science, and even artificial intelligence, influencing debates about the nature of truth, the limits of formal reasoning, and the role of intuition in mathematical discovery. These theorems continue to provoke deep reflection and research into the nature and scope of mathematical knowledge. This challenges the notion of mathematics as an absolute truth and underscores its nature as a construct.

The Debate: Discovery vs. Invention

The "discovery vs intention" debate in mathematics revolves around the philosophical question of whether mathematical concepts and truths are discovered as inherent aspects of an objective reality, or whether they are inventions created by human minds based on subjective intentions and conventions.

1. **Mathematics as Discovery:** Advocates of the "discovery" view argue that mathematical truths and structures exist independently of human thought. According to this perspective, mathematicians uncover pre-existing relationships and patterns that exist in the universe, much like discovering facts about the natural world. Proponents often

point to the universality and consistency of mathematical principles across cultures and time periods as evidence that mathematics reflects an objective reality waiting to be revealed.

2. **Mathematics as Invention:** On the other hand, proponents of the "invention" view assert that mathematics is a product of human creativity and intellect. In this view, mathematical concepts and systems are constructed by humans to organize and make sense of our observations and experiences. Mathematics, therefore, is seen as a tool or language devised by humans to describe patterns and relationships, rather than something inherent in the fabric of reality.

This debate is complex and multifaceted, touching on issues of epistemology (how knowledge is acquired), ontology (the nature of existence), and the philosophy of mathematics itself. Different mathematical schools of thought, such as formalism, intuitionism, and Platonism, offer varying perspectives on the nature of mathematical objects and their relationship to human cognition.

Ultimately, the discovery vs intention debate remains unresolved and continues to stimulate philosophical and mathematical inquiry. It underscores the profound and ambiguous relationship between mathematics and the human mind, challenging us to reconsider how we perceive and understand the nature of mathematical truth.

Or is it unresolved? While we do not know exactly when numbers and numerals were created, we can analyze the fact that the first true system of numbers, that we know today, was used as early as 3400 B.C. by the Sumerians. And that each major cultural group throughout time (Mayans, Aztecs, Romans, Egyptians, etc.) had their own numerals to represent numbers. Where did they get those from? There is no definitive answer, but the most realistic thing is that they made them up. It is all a human construct.

Chapter 4: Science - The Quest for Understanding

The Scientific Method

Science relies on the scientific method, a systematic process of observation, experimentation, and hypothesis testing.

Observation: The process begins with observing and noting a phenomenon or event in the natural world. This observation could arise from everyday experiences, previous research, or theoretical predictions.

Question: Based on the observation, scientists formulate a specific question or set of questions that they seek to answer through their investigation. This question should be clear, testable, and relevant to the phenomenon under study.

Hypothesis: A hypothesis is a proposed explanation for the observed phenomenon. It is a tentative statement that suggests a possible relationship between variables or factors. Hypotheses are typically based on prior knowledge, existing theories, or logical reasoning.

Prediction: From the hypothesis, scientists derive specific predictions or expectations about what will happen under certain conditions if the hypothesis is correct. These predictions guide the design of experiments or observations to test the hypothesis rigorously.

Experimentation or Observation: Scientists conduct controlled experiments or systematic observations to gather empirical evidence. Experiments involve manipulating variables under controlled conditions to test the predictions derived from the hypothesis. Observational studies involve gathering data from natural settings without altering the conditions.

Analysis: The data collected during the experiment or observation phase are analyzed using appropriate statistical or qualitative methods. This analysis helps determine whether the results support or refute the hypothesis and predictions.

Conclusion: Based on the analysis of the data, scientists draw conclusions regarding the validity of the hypothesis. If the results consistently support the hypothesis through repeated experimentation and observation, it may become accepted as a scientific theory—a well-substantiated explanation of some aspect of the natural world.

While it provides a robust framework for investigating the natural world, the scientific method itself is a construct, a human-devised approach to gaining knowledge.

Paradigms and Scientific Revolutions

Thomas Kuhn's concept of paradigms and scientific revolutions, introduced in his influential book "The Structure of Scientific Revolutions" (1962), significantly changed the way we understand the development of scientific knowledge. Here are the key ideas:

1. **Paradigms:** Kuhn defined paradigms as overarching frameworks or models that guide scientific inquiry within a particular field at a given time. Paradigms include not only theories and hypotheses but also the methods, standards, and assumptions that shape scientific research and interpretation. Paradigms provide a common set of concepts and practices that scientists

within a discipline use to understand and solve problems.

2. **Normal Science:** According to Kuhn, scientific research typically progresses through periods of "normal science" where scientists work within established paradigms. During these periods, scientists engage in puzzle-solving activities, testing and refining theories, and extending existing knowledge. Normal science aims to further develop and apply the paradigm rather than challenge its fundamental assumptions.

3. **Scientific Revolutions:** Kuhn argued that scientific progress is not gradual and cumulative but rather punctuated by periods of revolutionary change, known as scientific revolutions. These revolutions occur when anomalies or inconsistencies within the existing paradigm accumulate to the point where

the paradigm itself becomes increasingly unable to explain new observations or data.

4. **Crisis and Paradigm Shift:** The accumulation of anomalies leads to a crisis in the scientific community, where established theories and methods are questioned. This crisis can trigger a paradigm shift—a radical transformation in scientific thought where an old paradigm is replaced by a new one. Paradigm shifts involve a fundamental rethinking of scientific assumptions, theories, and methods, often accompanied by resistance from adherents of the old paradigm.

5. **Incommensurability:** Kuhn also introduced the concept of incommensurability, suggesting that paradigms are not merely different ways of looking at the same phenomena but

rather incompatible and difficult to compare directly. Different paradigms may use different concepts, methods, and standards, making it challenging to translate or communicate across paradigms.

Kuhn's work challenged the traditional view of science as a steady accumulation of knowledge and instead portrayed it as a dynamic process characterized by periods of stability (normal science) punctuated by revolutionary changes (scientific revolutions). His ideas sparked debates about the nature of scientific progress, the role of paradigms in shaping scientific inquiry, and the sociological aspects of scientific communities and their practices.

The Limits of Empiricism

Empiricism, the foundation of scientific inquiry, is based on sensory experience and observation.

However, sensory data is subject to interpretation and can be misleading. The limits of empiricism underscore the idea that our scientific knowledge is provisional and constructed from our perceptions.

Empiricism, as a philosophical approach, asserts that knowledge is primarily derived from sensory experience and observations. In the context of scientific inquiry, empiricism forms the foundation upon which hypotheses are tested, theories are formulated, and knowledge is validated. However, despite its foundational role, empiricism has several inherent limitations and challenges, particularly concerning the subjective nature of sensory experiences and observations.

1. **Subjectivity of Perception:** Sensory experiences, such as seeing, hearing, touching, and tasting, are inherently subjective. They depend not only on

external stimuli but also on individual perceptual mechanisms and cognitive processes. Variations in sensory acuity, prior experiences, cultural background, and psychological state can all influence how individuals perceive and interpret sensory information. This subjectivity introduces variability and potential biases into empirical observations, affecting the reliability and reproducibility of scientific findings.

2. **Theory-Ladenness:** The process of observation is not purely objective but is influenced by existing theories, assumptions, and conceptual frameworks—what philosophers of science refer to as theory-ladenness. Scientists interpret observations within the context of prevailing theories and paradigms, which can shape what is observed, how it is interpreted, and

which hypotheses are considered plausible. This interplay between theory and observation complicates the notion of pure, unbiased empirical inquiry.

3. **Measurement Limitations:** Empirical research often relies on measurements and instruments to quantify phenomena objectively. However, the accuracy and precision of these measurements can be limited by technological constraints, calibration errors, and the inherent complexity of the systems being studied. Additionally, choosing what and how to measure is influenced by theoretical assumptions and practical considerations, potentially introducing biases and limiting the scope of empirical investigations.

4. **Interpretation and Context Dependence:** Empirical observations require interpretation to make sense of

raw data in relation to scientific hypotheses. This interpretation is not straightforward and can be influenced by contextual factors, including cultural norms, disciplinary practices, and the goals of research. Different researchers may interpret the same empirical data differently, leading to divergent conclusions and interpretations within scientific communities.

5. **Inductive Reasoning and Generalization:** Empirical research often relies on inductive reasoning, where general conclusions are drawn from specific observations. However, the process of generalizing from limited observations to broader theories or laws is fraught with uncertainty. Inductive reasoning assumes that patterns observed in the past will continue to hold in the future, but this assumption is not

guaranteed and depends on the context and conditions under which observations were made.

While empiricism is a cornerstone of scientific inquiry, its reliance on sensory experiences and observations introduces inherent limitations due to the subjective nature of perception, the influence of pre-existing theories, measurement constraints, interpretative biases, and the challenges of inductive reasoning. Recognizing and addressing these limitations is crucial for advancing scientific knowledge and understanding the complexities of the natural world in a more nuanced and rigorous manner.

The Role of Theories and Models

The roles of theories and models in science are pivotal yet nuanced, serving as essential tools for understanding and navigating the complexities of the natural world. While they are indispensable for scientific inquiry, theories and models are not perfect or accurate representations of reality; rather, they are simplified constructs that aid human cognition in making sense of phenomena that exceed our immediate grasp.

1. **Theories as Frameworks:** Scientific theories are comprehensive explanations that integrate and organize a multitude of observations, experiments, and hypotheses into coherent frameworks. They provide overarching principles and laws that describe and predict natural phenomena, guiding further research and exploration. Theories are developed through iterative processes of observation, experimentation, and revision, aiming to capture essential aspects of reality while acknowledging their limitations and approximations.

2. **Models as Simplified Representations:**
    Models, on the other hand, are simplified
    representations of specific aspects or
    behaviors of natural systems. They can
    take various forms, including
    mathematical equations, computer
    simulations, diagrams, and physical
    replicas. Models are designed to capture
    essential features of a phenomenon while
    abstracting away unnecessary details,
    allowing scientists to explore hypotheses,
    predict outcomes, and test scenarios in a
    controlled manner.

3. **Abstraction and Simplification:** Both
    theories and models rely on abstraction
    and simplification to distill complex
    systems into manageable frameworks.
    This process involves identifying
    relevant variables, assumptions, and
    relationships that are essential for
    understanding the phenomena under

study. However, this simplification inevitably means that theories and models do not capture every nuance or detail of reality, leading to limitations in their predictive power and applicability in certain contexts.

4. **Human Cognitive Limits:** Theories and models are also shaped by human cognitive limitations and biases. Our brains evolved to understand immediate, tangible phenomena, often struggling with the intricacies of complex systems and phenomena that operate on different scales or under extreme conditions. Theories and models serve as cognitive tools that bridge this gap, enabling scientists to conceptualize and reason about phenomena beyond direct observation or intuitive understanding.

5. **Evolution and Revision:** Scientific theories and models are dynamic and

subject to evolution over time. They are refined, expanded, or even replaced as new evidence emerges, technologies advance, and conceptual frameworks evolve. This process reflects the ongoing pursuit of greater accuracy and explanatory power in scientific understanding, while acknowledging that our current models and theories are provisional and subject to revision.

In essence, while theories and models play indispensable roles in advancing scientific knowledge and understanding, they are simplified constructs that facilitate human comprehension of complex natural systems. They serve as pragmatic tools for organizing observations, making predictions, and guiding further investigation, yet they inherently fall short of capturing the full richness and intricacies of reality. Embracing the provisional

nature of theories and models encourages a
nuanced and critical approach to scientific
inquiry, fostering continuous refinement and
deeper insights into the natural world.

Chapter 5: Social Constructs and Cultural Realities

The Concept of Culture

Culture and beliefs are foundational aspects of human societies, shaping identities, behaviors, values, and interactions among individuals and groups. These concepts are intertwined and profoundly influence various dimensions of human life, including social norms, communication, worldview, and personal identity.

1. **Culture as a Complex System:** Culture can be understood as a complex system of shared beliefs, customs, traditions, values, symbols, and practices that are transmitted across generations within a society or community. It encompasses both tangible elements (such as language, rituals, and artifacts) and

intangible aspects (such as beliefs, norms, and worldviews). Culture provides a framework through which individuals make sense of their experiences, interpret the world around them, and navigate social interactions.

2. **Beliefs as Core Components:** Beliefs are central to culture, representing ideas, convictions, and assumptions held by individuals or groups about the nature of reality, morality, knowledge, and the supernatural. Beliefs can be religious, philosophical, political, scientific, or cultural in nature, influencing attitudes, behaviors, and decision-making processes. They provide a sense of identity and belonging within cultural contexts and often serve as guiding principles for individual and collective actions.

3. **Formation and Transmission:** Culture
   and beliefs are shaped through a dynamic
   process of formation, transmission, and
   adaptation. They evolve over time in
   response to historical, social, economic,
   and environmental factors. Cultural
   transmission occurs through various
   mechanisms, including formal education,
   family socialization, peer influence,
   media, and institutional practices. This
   transmission reinforces cultural norms
   and values while allowing for cultural
   diversity and innovation within societies.

4. **Diversity and Plurality:** Cultures and
   beliefs exhibit considerable diversity
   across different regions, ethnicities,
   religions, and social groups. This
   diversity enriches human experience and
   fosters tolerance, empathy, and
   understanding across cultural boundaries.
   However, cultural diversity can also lead

to misunderstandings, conflicts, and
challenges in achieving mutual respect
and cooperation in globalized societies.

5. **Adaptation and Change:** Cultures and
beliefs are not static but are adaptive and
responsive to changing social, political,
economic, and technological conditions.
Globalization, migration, urbanization,
and communication technologies
contribute to cultural exchange and
hybridization, influencing the evolution
of beliefs and cultural practices. As
societies interact and interconnect,
individuals may adopt, adapt, or
synthesize beliefs and cultural elements
from diverse sources, contributing to
cultural dynamism and innovation.

6. **Impact on Individuals and Societies:**
Culture and beliefs exert profound
influence on individual identity
formation, social cohesion, institutional

structures, and societal norms. They shape perceptions of self and others, influence decision-making processes, and provide frameworks for interpreting and responding to social challenges and opportunities. Understanding cultural diversity and respecting differing beliefs are essential for promoting inclusivity, social justice, and sustainable development in multicultural societies.

Culture also plays a fundamental role in defining shared realities and establishing societal norms within a community or society. It shapes perceptions, beliefs, values, behaviors, and institutions, providing a framework through which individuals understand themselves and interact with others. Here's how culture defines shared realities and societal norms:

1. **Shared Meanings and Interpretations:**
    Culture provides a shared system of
    meanings and interpretations that
    individuals within a society use to make
    sense of their experiences and the world
    around them. This includes language,
    symbols, rituals, myths, and narratives
    that convey collective beliefs, values,
    and historical perspectives. By
    internalizing these cultural elements,
    individuals develop a common
    understanding of reality and shared
    frameworks for interpreting events and
    behaviors.

2. **Socialization and Identity Formation:**
    From early childhood, individuals are
    socialized into their culture through
    family, education, peers, media, and
    religious institutions. Socialization
    instills cultural values, norms, and
    behavioral expectations, shaping

individual identity formation and influencing attitudes and behaviors throughout life. Cultural norms guide interactions, define acceptable behaviors, and reinforce social cohesion by establishing boundaries and expectations for social conduct.

3. **Preservation and Continuity:** Culture serves as a repository of collective knowledge, traditions, and practices passed down through generations. It preserves historical legacies, cultural heritage, and communal values, ensuring continuity and cohesion within societies over time. Cultural norms regarding family structure, gender roles, religious practices, and governance reflect historical experiences, social structures, and shared aspirations, guiding individuals in their roles and responsibilities within the community.

4. **Adaptation and Change:** While culture provides stability and continuity, it is also dynamic and adaptive to changing circumstances and external influences. Societal norms evolve in response to technological advancements, demographic shifts, globalization, and interactions with diverse cultures. Cultural adaptation involves integrating new ideas, practices, and beliefs while renegotiating existing norms to accommodate shifting social values and priorities. This process of cultural change can be gradual or abrupt, influenced by social movements, political shifts, economic developments, and generational transitions.

5. **Regulation of Behavior and Social Order:** Cultural norms and shared realities regulate individual and collective behavior by establishing

guidelines for acceptable conduct, social roles, and ethical standards. Norms provide a basis for resolving conflicts, enforcing social cohesion, and maintaining social order within communities. Violations of cultural norms may result in social sanctions, stigma, or legal consequences, underscoring the role of culture in shaping behavioral expectations and promoting social harmony.

6. **Cultural Diversity and Pluralism:** In multicultural societies, cultural diversity enriches shared realities by introducing alternative perspectives, values, and practices. Cultural pluralism acknowledges and respects the coexistence of multiple cultural identities and norms within a society, fostering tolerance, empathy, and mutual understanding among diverse groups.

Managing cultural diversity requires navigating tensions between preserving cultural traditions and promoting inclusive social norms that uphold human rights, equality, and justice for all individuals.

Social Norms and Values

Social norms and values are societal conventions and agreements within a culture or community that define acceptable behaviors, beliefs, and expectations. They shape interactions, relationships, and social structures by establishing standards for what is considered right or wrong, desirable or undesirable. However, these norms and values are constructs created by humans rather than objective truths or universal facts. Here's a deeper look into their nature and significance:

1. **Constructs of Human Society:** Social norms and values are constructs that evolve within specific cultural contexts and historical periods. They reflect collective beliefs, customs, and traditions that guide individuals in their interactions and decision-making processes. Norms may vary widely across cultures and subcultures, influencing everything from dress codes and manners to ethical principles and legal standards.

2. **Agreements and Consensus:** Norms and values emerge through a process of social agreement and consensus within a community. They are reinforced through socialization, education, and reinforcement mechanisms such as praise, reward, or social approval for conformity, and disapproval or sanctions for deviation. This consensus establishes

guidelines for behavior and helps
maintain social order and cohesion by
promoting predictability and mutual
understanding among members of a
society.

3. **Subjectivity and Cultural Relativity:**
   The concept of right or wrong, desirable
   or undesirable, is inherently subjective
   and culturally relative. What is
   considered appropriate behavior in one
   culture or context may be seen as
   inappropriate in another. Cultural
   relativism acknowledges that norms and
   values are shaped by cultural
   perspectives, historical experiences, and
   societal priorities, rather than being
   universally applicable truths.

4. **Fluidity and Change:** Social norms and
   values are not static but dynamic and
   adaptable to changing social, economic,
   and political conditions. They evolve

over time in response to shifting attitudes, values, and external influences. Social movements, technological advancements, globalization, and demographic shifts can challenge existing norms and promote new cultural norms that reflect evolving societal aspirations and challenges.

5. **Role in Social Control and Identity:** Norms and values play a crucial role in regulating individual and collective behavior, promoting social cohesion, and reinforcing group identity. They define roles and responsibilities within families, communities, and institutions, influencing how individuals perceive themselves and others. Norms can be formalized through laws and regulations or informally through customs and traditions, shaping ethical standards and expectations for moral conduct.

6. **Critical Reflection and Adaptation:**
While norms and values provide
structure and stability within societies,
they are also subject to critical reflection
and evaluation. Societies engage in
ongoing debates about the fairness,
relevance, and implications of existing
norms, leading to reforms, revolutions,
or cultural shifts that challenge
established conventions and promote
social justice, equality, and human rights.

In essence, social norms and values are
fundamental constructs of human societies that
define behavioral expectations, ethical
standards, and cultural identity. They are
agreements within communities rather than
absolute truths, reflecting shared understandings
and collective aspirations shaped by historical
contexts, cultural diversity, and ongoing social
dynamics. Recognizing the constructed nature

of norms encourages critical thinking and dialogue about societal priorities and the principles that guide human interaction and coexistence in a complex and interconnected world.

The Construct of Identity

Identity is a multifaceted concept that weaves together personal experiences, social interactions, and cultural affiliations. At its core, identity reflects not just who we are as individuals, but also how we perceive ourselves in relation to others and the world around us. However, what many fail to grasp is that identity is not merely an innate attribute—it is profoundly shaped by social constructs that categorize and define us within broader societal frameworks.

The Social Construction of Identity

Identity formation begins with the social constructs that delineate who we are perceived to be in society. These constructs include categories such as nationality, ethnicity, gender, and religion—labels that dictate how we are expected to behave, what roles we should play, and how we are viewed by others. These categories, while seemingly fixed, are far from immutable; they are human-made and subject to historical, cultural, and political influences.

Nationality: The concept of nationality assigns individuals to specific geopolitical entities, defining their legal and social standing within a nation-state. National identities are constructed through shared symbols, histories, and ideologies that bind individuals together under a common banner. Yet, nationality can be fluid, especially in contexts where migration, diaspora communities, and transnational identities challenge traditional notions of belonging.

Ethnicity: Ethnic identity revolves around shared cultural traits, ancestry, language, and heritage. Ethnic groups are often distinguished by common practices, customs, and traditions that reinforce a collective sense of belonging. However, ethnicity is not static; it evolves over time and can be reshaped by migration, cultural exchange, and intermarriage, blurring boundaries and challenging rigid definitions of identity.

Gender: Gender identity encompasses the social and cultural expectations associated with being male, female, or non-binary. These expectations dictate norms of behavior, appearance, and roles within society. Yet, gender is not purely biological; it is constructed through socialization processes that teach individuals how to perform gender according to societal norms. Gender identities are fluid and diverse, challenging binary understandings and advocating for inclusivity and self-expression.

Religion: Religious identity shapes beliefs, practices, and values that provide individuals with a moral and spiritual framework. Religious affiliations define communities and guide personal ethics, influencing social interactions and collective identities. However, religious identity can also be dynamic, as individuals navigate faith, spirituality, and secularism in a globalized world characterized by religious pluralism and multiculturalism.

Fluidity and Intersectionality

What becomes evident from examining these constructs is their fluidity and intersectionality—the ways in which multiple aspects of identity intersect and interact to shape individual experiences and perceptions. For instance, someone's national identity may intersect with their ethnic background, gender identity, and religious beliefs, creating a complex web of influences that mold their sense of self and social interactions.

Identity, therefore, is not a monolithic entity but a mosaic of interconnected factors that evolve over time and context. It is influenced by personal experiences, societal expectations, and historical legacies, highlighting the dynamic nature of human identity. Understanding identity as socially constructed encourages us to question rigid categories, challenge stereotypes, and embrace the diversity of human experiences and expressions.

In navigating the complexities of identity, individuals are empowered to explore and assert their authentic selves beyond societal norms and expectations. By recognizing the constructed nature of identity and the fluidity of its components, we embark on a journey of self-discovery and empathy, fostering a more inclusive and understanding society that celebrates the richness of human diversity.

Money and The Economy: Shared Fictions

Money, that elusive medium of exchange, holds a central place in human societies, yet its essence lies not in intrinsic value but in collective belief and societal constructs. As we delve into the complexities of money and the economy, we uncover a world built upon illusions and agreements that shape our understanding of wealth, value, and societal function.

The Social Construct of Money

Money is a symbolic representation of value, facilitating transactions and trade across diverse economic activities. Unlike tangible goods with inherent utility, money's worth is derived from shared trust and confidence within a community or society. This trust imbues money with purchasing power and enables it to serve as a store of wealth and a unit of account, essential for economic exchange and resource allocation.

Intrinsic Value: Contrary to commodities like food, shelter, or resources that fulfill basic human needs, money possesses no intrinsic value. Its worth is not rooted in physical properties or inherent utility but is instead conferred through collective belief and societal consensus. This belief underpins the stability and functionality of financial systems, where currencies fluctuate in value based on market perceptions and economic policies rather than inherent properties.

Medium of Exchange: Money's primary role as a medium of exchange facilitates the efficient allocation of resources and goods within economies. It enables specialization, trade, and economic growth by reducing the inefficiencies of barter systems and enabling transactions across vast geographical and social distances. Yet, its effectiveness depends on widespread acceptance and trust in its reliability and stability as a means of exchange.

The Constructed Nature of Financial Systems

Beyond money, the entire economic system is constructed upon agreements, norms, and institutions that govern production, distribution, and consumption. The stock market, for instance, exemplifies this constructed reality—where financial instruments, securities, and investments derive their value from market perceptions, investor confidence, and regulatory frameworks.

Stock Market Dynamics: The stock market, a cornerstone of modern economies, exemplifies the constructed nature of financial systems. Here, stocks and bonds represent ownership stakes and debt obligations in corporations, traded based on investor expectations of future profitability and economic conditions. Market fluctuations reflect collective sentiments, speculation, and external factors rather than the intrinsic values of companies or assets.

Illusions of Wealth: The pursuit of wealth and financial success often revolves around illusions of security and prosperity fostered by monetary accumulation. Wealth, measured in currency or assets, symbolizes status, power, and opportunity within societal hierarchies. Yet, this pursuit can perpetuate inequalities and economic disparities, as financial outcomes are influenced by factors beyond individual effort or merit.

Revisiting the Construct

In acknowledging the constructed nature of money and economy, we confront the paradox of an economic system built upon intangible beliefs and agreements. The resilience and functionality of financial systems hinge on trust, transparency, and regulatory frameworks that uphold economic stability and fairness. However, periodic crises and disruptions—such as market crashes or economic recessions—underscore the fragility and interconnectedness of global economies.

Turning it On: Despite its flaws and inherent vulnerabilities, the reactivation of economic systems, like the stock market, after disruption (the Great Depression and the 2007-2008 financial crisis) underscores their fundamental role in societal functioning and human livelihoods. The decision to restore economic activity is futile. Why do we keep turning the stock market back on if all it has been proven to do is keep the poor poor and keep the rich rich?

Money and the economy are powerful social constructs that shape human behavior, societal structures, and global interactions. They represent a delicate balance of trust, belief, and institutional frameworks that mediate economic activities and resource allocation. By understanding their constructed nature, we navigate the complexities of wealth, value, and economic exchange with an awareness of their inherent illusions and the collective responsibilities they entail.

Amidst the constructed nature of money and economy lies a profound realization: while these systems shape our societal structures and daily lives, they do not define our happiness or fulfillment. As we navigate the complexities of economic existence, we are reminded of the importance of authenticity, purpose, and genuine human connections in shaping our experiences and priorities.

Beyond Material Accumulation

In a world driven by consumerism and material wealth, the pursuit of happiness often becomes intertwined with the accumulation of possessions and financial success. However, true fulfillment transcends materialistic endeavors and lies in aligning our actions with personal values, passions, and meaningful relationships. Choosing happiness involves:

1. Embracing Authenticity: Authenticity entails living in alignment with our true selves—our beliefs, aspirations, and ethical principles—rather than conforming to societal expectations or external pressures. It involves introspection and self-discovery to uncover what truly brings joy and purpose to our lives, beyond superficial measures of success or status.

2. Prioritizing Relationships: Human connections and social bonds are essential sources of happiness and well-being. Investing time and effort in nurturing meaningful relationships with family, friends, and communities fosters a sense of belonging, support, and emotional fulfillment that transcends material wealth.

3. Cultivating Inner Fulfillment: Inner fulfillment arises from personal growth, self-acceptance, and the pursuit of passions and interests that bring intrinsic joy and satisfaction. Whether through creative expression, lifelong learning, or contributing to causes larger than ourselves, cultivating inner fulfillment enriches our lives and contributes to a sense of purpose beyond material gain.

Conscious Living and Well-being

Conscious living involves mindful choices and intentional actions that prioritize well-being and sustainability over relentless pursuit of material possessions or financial gain. It encompasses:

1. Sustainable Consumption: Adopting sustainable lifestyles and consumption habits that minimize environmental impact and promote ethical practices in production and consumption. Conscious consumerism involves making informed choices that align with values of environmental stewardship and social responsibility.

2. Mental and Emotional Health: Prioritizing mental and emotional well-being through practices such as mindfulness, self-care, and stress management. Cultivating resilience and emotional intelligence equips individuals to navigate challenges and setbacks with grace and perspective, fostering greater overall happiness and life satisfaction.

3. Contribution and Impact: Finding fulfillment through contributions to community, society, or causes that resonate deeply. Whether through volunteering, advocacy, or entrepreneurship that serves social good, making a positive impact fosters a sense of purpose and fulfillment beyond personal gain.

Embracing Freedom and Choice

While the constructs of money and economy exert influence over societal structures and economic activities, individuals possess agency in shaping their personal narratives and defining success on their own terms. Embracing freedom involves:

1. Redefining Success: Challenging conventional notions of success and redefining personal goals and aspirations based on intrinsic values, passions, and desired impact. Success becomes a journey of self-discovery and growth rather than a destination dictated by external validations.

2. Resilience in Adversity: Building resilience and adaptability to navigate uncertainties and economic fluctuations with courage and determination. Embracing setbacks as opportunities for learning and growth fosters resilience and strengthens personal resolve in pursuing happiness and fulfillment.

3. Choosing Happiness: Ultimately, choosing happiness involves embracing gratitude for the present moment, fostering meaningful connections, and finding joy in simple pleasures and experiences. It is a conscious decision to prioritize well-being, authenticity, and personal fulfillment over the pursuit of external validations or materialistic pursuits.

In embracing authenticity, conscious living, and the pursuit of happiness beyond material accumulation, individuals reclaim agency in defining their own narratives and shaping a life that resonates with purpose, meaning, and genuine fulfillment. By navigating the constructed realities of money and economy with awareness and intentionality, we empower ourselves to lead lives rich in happiness, connection, and personal fulfillment.

Chapter 6: The Constructs of Time and Space

The Nature of Time

Time, the intangible thread weaving through the fabric of human existence serves as a fundamental construct shaping how we perceive, organize, and make sense of the world. From ancient civilizations to modern societies, the concept of time has evolved into a multifaceted framework that influences cultural practices, societal structures, and individual experiences. This chapter explores the nature of time as a human construct, its role in human perception, and the diverse cultural interpretations that highlight its subjective and varied nature.

The Conceptualization of Time

At its essence, time represents the progression of events from the past, through the present, to the future—a linear framework that provides a sense of continuity and order to human experience. However, beyond its objective measurement through clocks and calendars, time is fundamentally shaped by subjective interpretations, cultural norms, and societal conventions that vary widely across different civilizations and historical epochs.

1. Human Perception and Sense-Making:
Time serves as a cognitive tool for organizing experiences, memories, and expectations. It allows individuals to anticipate future events, learn from past occurrences, and navigate the complexities of daily life. This subjective experience of time influences decision-making, personal narratives, and the construction of

individual identities within temporal
frameworks.

2. Cultural Variability: Across cultures,
perceptions and structures of time exhibit
remarkable diversity, reflecting unique
cosmological beliefs, historical contexts, and
social practices. For instance:

- **Linear vs. Cyclical Time:** Western
  cultures often adopt a linear view of
  time, where history unfolds along a
  chronological continuum from past to
  present to future. In contrast, many
  indigenous cultures and Eastern
  philosophies embrace cyclical
  conceptions of time, where events recur
  in cycles of seasons, generations, or
  cosmic epochs.

- **Event-Oriented vs. Clock-Oriented
  Time:** Some societies prioritize
  punctuality and adherence to precise

schedules (clock-oriented time),
reflecting industrialized economies and
urban lifestyles. In contrast,
event-oriented time emphasizes the
fluidity of events and communal
rhythms, as seen in agrarian societies or
traditional cultures guided by natural
phenomena.

- **Temporal Fluidity:** Cultural perceptions
of time may also manifest in temporal
fluidity, where past, present, and future
coexist in dynamic interaction. This
perspective values the
interconnectedness of temporal
dimensions and the continuity of
ancestral legacies within contemporary
practices.

Time as Social Construct

The social construction of time extends beyond
individual perception to encompass collective

agreements, institutional frameworks, and cultural rituals that regulate temporal activities and societal rhythms. These constructs include:

1. Calendars and Timekeeping Systems: Calendars, whether lunar, solar, or lunisolar, structure temporal cycles and mark significant cultural events, religious observances, and agricultural practices. Timekeeping devices—from ancient sundials to modern atomic clocks—standardize temporal measurements and synchronize societal activities.

2. Rituals and Commemorations: Rituals, ceremonies, and commemorations embed cultural meanings within temporal milestones, reinforcing collective memory and identity. These temporal markers—from birthdays and anniversaries to seasonal festivals and rites of passage—celebrate continuity, renewal, and community cohesion across generations.

3. Temporal Hierarchies: Temporal hierarchies, such as historical epochs, eras, and generations, contextualize human experiences within broader narratives of change, progress, and societal evolution. These temporal constructs shape historical consciousness, collective identities, and interpretations of past achievements and challenges.

Embracing Temporal Diversity

The diversity of temporal perceptions and cultural practices underscores the subjective and constructed nature of time as a human phenomenon. By exploring diverse cultural interpretations of time, individuals gain insights into alternative frameworks for understanding existence, temporality, and the interconnectedness of human experience across global landscapes.

Time emerges as a dynamic and culturally contingent construct that transcends mere measurement to encompass profound implications for human cognition, social organization, and cultural identity. By embracing temporal diversity and recognizing the fluidity of temporal constructs, individuals navigate the complexities of time with curiosity, empathy, and a deeper appreciation for the rich tapestry of human existence across past, present, and future horizons.

Expanding on Calenders And How They Vary

Different cultures around the world have developed diverse calendar systems to organize time, mark significant events, and regulate social, religious, and agricultural activities. These calendars vary in structure, methodology, and cultural significance, reflecting unique cosmological beliefs, historical developments, and societal needs. Here are some examples of how different cultures have set up their calendars:

Gregorian Calendar (Western Calendar)

The Gregorian calendar is the most widely used civil calendar globally today. It was introduced by Pope Gregory XIII in 1582 to reform the Julian calendar, which had become out of sync with the solar year. Key features include:

- **Solar Calendar:** Based on the solar year, with 365 days divided into 12 months. It incorporates leap years every four years (except for years divisible by 100 but not by 400) to account for the extra fraction of a day in the solar year.
- **Months:** Named after Roman deities or numbers (January, February, etc.).

- **Weekdays:** Seven days a week, named after Roman gods (Sunday, Monday, etc.).

Islamic Calendar (Hijri Calendar)

The Islamic calendar is a lunar calendar used by Muslims worldwide to determine religious observances. It follows the cycles of the moon rather than the solar year:

- **Lunar Calendar:** Consists of 12 lunar months of 29 or 30 days each, totaling about 354 or 355 days in a year.
- **Months:** Begin with the sighting of the new moon, with months such as Ramadan and Dhul-Hijjah holding special religious significance.
- **No Leap Years:** The Islamic calendar does not have leap years, causing it to slowly drift relative to the Gregorian calendar.

Chinese Calendar

The Chinese calendar, also known as the Agricultural Calendar, has been used in China for millennia. It combines lunar and solar elements:

- **Lunisolar Calendar:** Based on both lunar phases and solar terms to mark agricultural seasons.
- **12-Year Cycle:** Each year is associated with one of 12 animal zodiac signs (Rat, Ox, Tiger, etc.).
- **24 Solar Terms:** Divides the solar year into 24 periods based on the sun's position relative to the ecliptic.

Hindu Calendar (Vedic Calendar)

The Hindu calendar is a complex system used in India for religious and cultural purposes:

- **Lunisolar Calendar:** Combines lunar months with solar years.

- **Regional Variations:** Different regions may use different calculations and variations of the calendar.

- **Months:** Named after astronomical events and deities, with religious festivals and rituals tied to specific lunar phases.

Hebrew Calendar

The Hebrew calendar is a lunisolar calendar used for Jewish religious observances:

- **Lunisolar Calendar:** Months are based on lunar cycles, with adjustments to synchronize with the solar year.

- **Leap Years:** Occur in a 19-year cycle to align lunar months with solar years.

- **Months:** Named in Hebrew, with special religious significance attached to holidays such as Passover and Yom Kippur.

Maya Calendar

The ancient Maya civilization in Mesoamerica developed several calendars, including the Long Count and the Tzolk'in:

- **Long Count:** A linear count of days since a mythical starting point, used for historical events and celestial cycles.
- **Tzolk'in:** A 260-day ritual calendar consisting of 20 periods of 13 days each, combined with a 365-day solar calendar.

Which one is correct? There is no correct calendar. It was all constructed by humans to make sense of the passing of time and to make these insanely complex things make sense to us. Because the truth is, no one knows. Technically, there is no truth. There is no definitive answer for anything.

The Construct of Space

Space, the expansive canvas upon which the universe unfolds, is a concept deeply embedded in human understanding and exploration. From ancient cosmologies to modern scientific theories, the concept of space has evolved, revealing its complexities and challenging conventional constructs. This chapter explores the nature of space as a foundational concept in human cognition, the insights from scientific advancements, and the enduring mysteries that remind us of our limited understanding.

Foundations of Human Understanding

Space serves as the framework within which we perceive and navigate the world. It encompasses the three dimensions of length, width, and height, providing a context for spatial relationships, distances, and physical boundaries. Our everyday experiences are shaped by spatial awareness, from perceiving the proximity of objects to navigating environments with spatial reasoning.

1. Perception and Navigation: Human cognition relies on spatial awareness to interpret surroundings, coordinate movements, and interact with our environment. Spatial perception enables us to navigate cities, explore landscapes, and conceptualize distances, influencing how we perceive scale, direction, and orientation.

2. Cultural and Artistic Representations: Throughout history, diverse cultures have conceptualized space through art, architecture, and cosmological beliefs. Spatial representations in art and literature reflect cultural values, societal structures, and interpretations of cosmic order, demonstrating the interplay between imagination and spatial perception.

Scientific Insights and Challenges

Scientific advancements have revolutionized our understanding of space, revealing its dynamic and non-absolute nature. The theories of relativity and quantum mechanics have challenged conventional notions of space as a static backdrop, introducing new dimensions of complexity and uncertainty.

1. Relativity Theory: Albert Einstein's theory of General Relativity revolutionized our understanding of space and time as interwoven dimensions, where gravitational forces shape the curvature of spacetime. This conceptual framework explains phenomena such as black holes, gravitational waves, and the bending of light, demonstrating the fluidity and dynamic nature of spacetime.

2. Quantum Mechanics: At the quantum scale, the principles of quantum mechanics defy classical intuitions about space and matter. Quantum entanglement suggests that particles can be interconnected regardless of distance, challenging notions of locality and separability in space. Quantum field theory further explores the concept of space as a quantum vacuum filled with fluctuating fields and virtual particles, highlighting the elusive and probabilistic nature of spacetime.

Limits of Human Knowledge

Despite scientific progress, space remains a frontier of exploration and inquiry, marked by unanswered questions and enduring mysteries:

1. Dark Matter and Dark Energy: A significant portion of the universe consists of dark matter and dark energy, entities whose nature and properties remain elusive. They influence cosmic structures and expansion rates, yet their exact composition and interaction with spacetime are still unknown.

2. Multiverse Hypotheses: Theoretical frameworks, such as the multiverse theory, propose the existence of multiple universes with varying physical laws and dimensions of space. These hypotheses challenge our understanding of space as a singular entity and raise profound questions about the nature of existence and cosmic diversity.

Embracing the Unknown

In exploring the construct of space, we confront the limitations of human knowledge and the boundless possibilities for discovery. The evolving perspectives from cultural interpretations to scientific theories underscore the fluidity and complexity of space as a conceptual framework. As we navigate the mysteries of the cosmos, we embrace humility and curiosity, recognizing that our understanding of space continues to evolve alongside scientific inquiry and human imagination.

Space remains a foundational concept in human cognition and scientific exploration, shaping our understanding of the universe's vastness and complexity. From cultural representations to scientific revelations, the construct of space invites us to contemplate the boundaries of human knowledge and the enduring quest to unravel the mysteries of cosmic existence. And you're probably thinking, "Space isn't a concept, it is real and exists." But how do we know that is outer space? How do we know that's exactly what it is? That's what it is to us. But that doesn't mean anything. It does not mean that space is space. We know nothing. I know nothing. You know nothing. Nothing is factual. Nothing is correct because we don't know.

Relativity and Quantum Mechanics

Einstein's theory of relativity comprises two major theories: Special Relativity and General Relativity, both of which revolutionized our understanding of space, time, and gravity.

1. **Special Relativity:**
   - **Principles:** Introduced in 1905, Special Relativity proposes that the laws of physics are the same for all observers in uniform motion (i.e., moving at a constant velocity).
   - **Key Postulates:**

- **Constancy of the Speed of Light:** The speed of light in a vacuum is constant and is the same for all observers, regardless of their motion relative to the light source.
  - **Relativity of Simultaneity:** Events that are simultaneous for one observer may not be simultaneous for another moving observer.
- **Implications:** Special Relativity fundamentally altered our concepts of space and time. It unified space and time into a single entity called spacetime, where measurements of distances and durations depend on the

observer's motion relative to the observed objects.

2. **General Relativity:**

    - **Principles:** Formulated in 1915, General Relativity extends Special Relativity to include the effects of gravity.

    - **Key Concepts:**

        - **Curvature of Spacetime:** Gravity is not a force but a curvature of spacetime caused by the presence of mass and energy. Massive objects like planets and stars curve spacetime around them, causing other objects to move along curved paths.

        - **Equivalence Principle:** The effects of gravity experienced by an observer in a gravitational field are indistinguishable from those experienced by an observer in an accelerated reference frame.

- **Implications:** General Relativity has provided accurate predictions for various astronomical phenomena, such as the bending of light by gravity (gravitational lensing) and the existence of black holes.

Quantum Mechanics:

Quantum mechanics is the branch of physics that describes the behavior of particles at microscopic scales, challenging classical physics and introducing probabilistic principles.

1. **Wave-Particle Duality:**

- **Principles:** Quantum mechanics proposes that particles such as electrons and photons exhibit both wave-like and particle-like behaviors.

- **Quantization:** Certain physical quantities, such as energy levels in atoms, are quantized—they can only take on discrete values.

- **Uncertainty Principle:** Formulated by Werner Heisenberg, this principle states that certain pairs of physical properties, such as position and momentum, cannot be precisely determined simultaneously. There is an inherent uncertainty in their measurement.

2. **Quantum Superposition and Entanglement:**

   - **Superposition:** Quantum particles can exist in multiple states simultaneously until measured, a phenomenon known as superposition.

   - **Entanglement:** Quantum entanglement occurs when two or more particles become correlated in such a way that the state of one particle instantaneously affects the state of another, even if they are separated by vast distances.

3. **Applications and Challenges:**

   - **Technological Advances:** Quantum mechanics has led to the development of technologies such as transistors, lasers, and MRI machines.

- o **Challenges:** Quantum mechanics challenges our intuitive understanding of reality, introducing concepts that defy classical logic and everyday experience, such as particles existing in multiple states or being interconnected regardless of distance.

Einstein's theory of relativity and quantum mechanics represent profound advancements in our understanding of the universe, yet they are theoretical frameworks rather than absolute truths. Here's an exploration of how these theories suggest the relativity and interconnected nature of time and space, and their limitations in the broader context of our understanding:

Relativity and Interconnectedness

1. Einstein's Theory of Relativity:

- **Concepts:** Special and General
  Relativity propose that space and time
  are intertwined in a four-dimensional
  spacetime continuum. They suggest that
  measurements of time and space can
  vary depending on an observer's motion
  and gravitational field.
- **Implications:** This challenges the
  classical Newtonian view where space
  and time were considered separate and
  absolute. Instead, relativity posits that
  these dimensions are dynamic and
  influenced by the presence of mass and
  energy.

2. Quantum Mechanics:

- **Principles:** Quantum mechanics
  introduces probabilistic principles and

the wave-particle duality, suggesting that particles do not have definite properties until measured.

- **Implications:** Quantum entanglement and superposition indicate that particles can influence each other instantaneously across vast distances, defying classical notions of locality and causality.

Theoretical Constructs

1. Human-Centric Perspective:

- **Observational Bias:** Both relativity and quantum mechanics are based on observations and experiments conducted within the human-scale universe. They provide accurate predictions and explanations within this framework but may not be directly applicable to extremes of scale, such as the very small (quantum realm) or the very large (cosmological scales).

2. Theoretical Boundaries:

- **Limits of Application:** Relativity and quantum mechanics operate within their respective domains but are not fully unified into a single theory that explains all phenomena. Theories like quantum gravity seek to bridge these gaps but remain speculative.

Philosophical Considerations

1. Subjective Understanding: The theories of relativity and quantum mechanics highlight the role of observation and perspective in shaping our understanding of the universe. They emphasize that scientific models are tools for human comprehension rather than definitive descriptions of reality.

2. Continuing Exploration: Scientific inquiry continues to probe the frontiers of time, space, and fundamental forces. The search for a unified theory that reconciles quantum mechanics with gravity (quantum gravity) reflects ongoing efforts to refine and expand our understanding beyond current theoretical frameworks.

While Einstein's theory of relativity and quantum mechanics have revolutionized physics and expanded our comprehension of time, space, and the universe, they remain theoretical constructs subject to refinement and interpretation. They provide powerful tools for making predictions and understanding natural phenomena within human-scale experiences but acknowledge the limits of our current observational and theoretical frameworks. Ultimately, these theories underscore the dynamic and evolving nature of scientific understanding, encouraging ongoing exploration and critical inquiry into the fundamental nature of reality.

The Human Perception of Time and Space

Human perception of time and space is a complex interplay of cognitive processes, cultural influences, and subjective experiences. Our understanding of these fundamental dimensions is shaped by both biological factors that govern sensory input and cognitive processing, as well as cultural frameworks that dictate how we interpret and organize these experiences.

Barely Scratching The Surface

Our perception of everything. Every color we see, every tree we touch, every car we drive, and every person we talk to is only .0035% of reality. The visible spectrum, the range of electromagnetic radiation that the human eye can detect, is a tiny fraction of the vast electromagnetic spectrum. The colors we perceive in a rainbow—from red to violet—are merely a sliver of the multitude of wavelengths that exist. This chapter delves into the nature of the electromagnetic spectrum, the limitations of human perception, and the implications of these limitations for our understanding of reality.

The Electromagnetic Spectrum

1. Understanding the Spectrum:

- **Range of Wavelengths:** The electromagnetic spectrum encompasses all types of electromagnetic radiation, from the longest radio waves to the shortest gamma rays. It includes radio waves, microwaves, infrared, visible light, ultraviolet, X-rays, and gamma rays.

- **Visible Light:** The visible spectrum ranges approximately from 380 to 750 nanometers in wavelength. This range represents only about 0.0035 percent of the entire electromagnetic spectrum.

2. Beyond Human Vision:

- **Infrared and Ultraviolet:** Just beyond the visible spectrum are infrared and ultraviolet light. Infrared is longer in wavelength than visible light and is associated with heat. Ultraviolet light has shorter wavelengths than visible light and can cause sunburn.
- **X-rays and Gamma Rays:** Further along the spectrum, X-rays and gamma rays have even shorter wavelengths and higher energies, used in medical imaging and cancer treatment.

Limitations of Human Perception

1. Biological Constraints:

- **Eye Structure:** Human eyes are equipped with photoreceptor cells—rods and cones—that detect specific ranges of wavelengths. Cones are responsible for color vision and are sensitive to red, green, and blue light.
- **Color Perception:** The combination of signals from different types of cones allows humans to perceive a wide array of colors, but this perception is limited to the visible spectrum.

2. Perceptual Implications:

- **Invisible Realities:** Many phenomena occur outside the visible spectrum, from the heat emitted by objects (infrared) to the high-energy processes in the universe (X-rays and gamma rays). These are imperceptible to the human eye without technological assistance.

- **Technological Enhancements:** Instruments like infrared cameras, ultraviolet sensors, and X-ray machines extend our sensory capabilities, allowing us to observe and study these otherwise invisible aspects of reality.

Cognitive Foundations

1. Sensory Input and Brain Processing:

- **Temporal and Spatial Processing:** The human brain receives sensory inputs from the environment, including visual, auditory, tactile, and proprioceptive cues. These inputs are processed in specialized regions of the brain that handle temporal (time-related) and spatial (location-related) information.

- **Integration and Perception:** Neural networks integrate sensory data to construct a coherent perception of time and space. This integration involves complex computations that synthesize incoming information into a unified experience of reality.

2. Illusion of Continuity:

- **Temporal Binding:** Our brains create a sense of continuity by binding discrete sensory events into a seamless stream of experience. This temporal binding allows us to perceive events as unfolding in a continuous flow, despite the discrete nature of sensory inputs.
- **Spatial Awareness:** Similarly, spatial awareness involves integrating visual and proprioceptive information to navigate and interact with our surroundings. Spatial perception allows

us to orient ourselves in space, estimate distances, and interact with objects effectively.

Cultural Influences

1. Cultural Frameworks:

- **Concepts of Time:** Different cultures conceptualize and organize time in varied ways. Some cultures emphasize punctuality and adherence to schedules, while others prioritize event-oriented time where activities unfold in response to natural or social cues.

- **Spatial Representations:** Cultural perspectives influence how space is perceived and organized. Cultural norms dictate spatial relationships, architectural styles, and navigation techniques that reflect societal values and historical contexts.

2. Language and Thought:

- **Linguistic Relativity:** The Sapir-Whorf hypothesis suggests that language shapes thought processes and perception. Different languages encode temporal and spatial concepts differently, influencing how individuals perceive and reason about time and space.

- **Cultural Practices:** Rituals, ceremonies, and daily practices embed temporal and spatial meanings within cultural contexts. These practices reinforce collective identity, memory, and social cohesion through shared temporal and spatial frameworks.

Subjective Experience

1. Subjective Time Perception:

- **Psychological Time:** Time perception is subjective and can vary based on emotional state, attentional focus, and contextual factors. Time can feel elongated or contracted depending on subjective experience, such as during moments of excitement or boredom.

- **Temporal Biases:** Cognitive biases, such as hindsight bias or the tendency to underestimate future events (prospect theory), further shape our perception of past, present, and future.

2. Spatial Orientation:

- **Personal Space:** Individual differences in spatial perception include preferences for personal space, spatial navigation abilities, and cultural norms regarding physical proximity and social interaction.

- **Environmental Influence:** Physical environments influence spatial perception, with urban versus rural settings offering distinct spatial cues and navigational challenges.

Human perception of time and space is a dynamic interplay of biological processes, cultural frameworks, and subjective experiences. Our brains construct a coherent reality from sensory inputs, creating the illusion of a stable and continuous temporal-spatial environment. By understanding the cognitive foundations and cultural influences that shape our perception, we gain insight into the diverse ways individuals and societies navigate and interpret the dimensions of time and space. This exploration invites us to reflect on the complex interconnections between perception, cognition, and culture, illuminating the rich tapestry of human experience in our ever-evolving understanding of reality.

Chapter 7: Morality and Ethics - Human-Made Guidelines

The Origins of Moral Thought

Moral thought, the foundation of ethical reasoning and societal norms, has evolved over millennia, shaped by diverse cultural, philosophical, and psychological influences. This chapter delves into the origins of moral thought, exploring its emergence in human societies, the role of evolutionary processes, and the philosophical inquiries that continue to shape our understanding of right and wrong.

Evolutionary Foundations

1. Evolutionary Psychology:

- **Social Behavior:** Evolutionary theories propose that moral sentiments, such as empathy and cooperation, evolved as adaptive traits to promote social cohesion and survival. Early humans who cooperated and cared for others were more likely to thrive and pass on their genes.

- **Reciprocal Altruism:** The concept of reciprocal altruism suggests that individuals engage in moral behaviors, such as sharing resources or helping others, with the expectation of future benefits or reciprocation.

2. Moral Emotions:

- **Empathy and Compassion:** Moral thought is intertwined with emotions like empathy, which allows individuals to recognize and respond to the suffering or well-being of others. Compassion motivates altruistic actions and forms the basis of moral judgments.
- **Guilt and Shame:** Moral emotions like guilt and shame emerge from social

interactions and serve to regulate
behavior within societal norms. These
emotions reinforce moral guidelines and
encourage adherence to ethical standards.

Cultural Development

1. Cultural Relativism:

- **Variability of Moral Norms:** Different
  cultures exhibit diverse moral norms and
  values shaped by historical,
  geographical, and religious influences.
  Cultural relativism acknowledges that
  moral judgments are context-dependent
  and vary across societies.
- **Cultural Transmission:** Moral norms
  are transmitted through cultural
  practices, rituals, and teachings,
  reinforcing ethical frameworks and
  shaping individual moral identities.

2. Philosophical Inquiries:

- **Ethical Theories:** Philosophical traditions, from ancient Greece to modern times, have explored ethical dilemmas and principles. Utilitarianism, deontology, virtue ethics, and consequentialism offer distinct perspectives on the foundations of moral thought and decision-making.

- **Metaethics:** Metaethical inquiries examine the nature of moral judgments, exploring questions about the objectivity or subjectivity of ethical norms and the possibility of moral knowledge.

Cognitive and Developmental Perspectives

1. Moral Development:

- **Piaget and Kohlberg:** Psychologists like Jean Piaget and Lawrence Kohlberg

proposed theories of moral development,
suggesting that individuals progress
through stages of moral reasoning from
childhood to adulthood. This
development involves understanding
rules, empathy, and abstract ethical
principles.

- **Social Learning:** Children learn moral
values through observation,
reinforcement, and socialization within
families, schools, and communities.

2. Moral Decision-Making:

- **Cognitive Processes:** Moral judgments
involve cognitive processes, including
moral reasoning and intuitive responses.
Dual-process theories propose that moral
decisions arise from both deliberate
reasoning and automatic emotional
responses.

- **Moral Dilemmas:** Ethical dilemmas challenge individuals to weigh competing moral principles or values, requiring reflection and ethical decision-making strategies.

Contemporary Challenges and Reflections

1. Ethical Diversity:

- **Globalization:** In a globalized world, ethical dilemmas arise from cultural diversity and conflicting moral values. Debates over human rights, environmental ethics, and technological advancements highlight the need for cross-cultural dialogue and ethical reflection.
- **Emerging Issues:** Contemporary issues such as artificial intelligence ethics,

genetic engineering, and climate change ethics raise complex moral questions that require interdisciplinary approaches and ethical frameworks.

2. Moral Progress:

- **Social Movements:** Throughout history, social movements have challenged prevailing moral norms and advocated for justice, equality, and human rights. Moral progress involves societal shifts toward more inclusive, compassionate, and equitable ethical standards.

- **Future Directions:** The evolution of moral thought continues to unfold through ongoing philosophical inquiry, scientific research, and societal debates. Ethicists and scholars explore new frontiers in bioethics, digital ethics, and social justice to address emerging ethical challenges.

The origins of moral thought encompass evolutionary, cultural, philosophical, and psychological dimensions that collectively shape human understanding of ethical principles and values. From evolutionary adaptations to philosophical reflections and societal developments, moral thought reflects humanity's quest for fairness, compassion, and ethical integrity. By examining the diverse origins and contemporary challenges of moral thought, we deepen our appreciation for the complexities of ethical reasoning and the ongoing evolution of moral norms in our shared human experience.

Ethical Theories and Constructs

Ethical theories provide frameworks for understanding and evaluating moral principles, guiding individuals and societies in determining right and wrong. From utilitarianism to deontology and virtue ethics, these constructs offer distinct perspectives on ethics, reflecting human-made attempts to address ethical dilemmas and moral judgments.

Utilitarianism

1. Principle of Utility:

- **Foundations:** Utilitarianism, proposed by philosophers like Jeremy Bentham and John Stuart Mill, asserts that actions are morally right if they maximize happiness or well-being (utility) for the greatest number of people.
- **Consequentialism:** Utilitarianism is a form of consequentialism, where the morality of an action is judged by its outcomes or consequences rather than inherent qualities.
- **Critiques:** Critics argue that utilitarianism can overlook individual rights or justice concerns in favor of

overall utility, leading to potential injustices or sacrificing minority interests for the majority.

Deontology

1. Moral Duties and Rights:

- **Immanuel Kant's Categorical Imperative:** Deontology emphasizes moral duties and obligations based on principles of reason and universalizability. Kant's categorical imperative states that actions should be guided by principles that could be universally applied without contradiction.

- **Non-Consequentialist Ethics:** Deontology contrasts with consequentialist theories by prioritizing the inherent rightness or wrongness of actions rather than their outcomes.

- **Critiques:** Critics argue that deontological principles may not provide clear guidance in complex moral dilemmas or conflicts between duties.

Virtue Ethics

1. Character and Virtue:

- **Aristotle's Virtue Ethics:** Virtue ethics focuses on cultivating moral character and virtues (e.g., courage, honesty) as the basis for ethical behavior. Aristotle emphasized the development of virtues through habituation and moral education.
- **Agent-Centered Ethics:** Unlike utilitarianism and deontology, virtue ethics centers on the moral character of the agent rather than specific actions or consequences.
- **Critiques:** Critics argue that virtue ethics may lack universal principles and clear

guidelines for resolving moral conflicts
or dilemmas.

Constructed Nature of Ethical Theories

1. Human-Made Frameworks:

- **Historical Development:** Ethical
  theories evolve in response to
  philosophical inquiries, cultural norms,
  and societal challenges. They reflect
  human attempts to rationalize and
  systematize moral judgments and
  principles.

- **Subjectivity and Contingency:** Ethical
  theories are shaped by cultural contexts,
  historical perspectives, and individual
  interpretations of morality. They offer
  frameworks for ethical deliberation but
  are not universally applicable or
  objective truths.

2. Interdisciplinary Perspectives:

- **Philosophical Inquiry:** Ethicists and philosophers continue to debate and refine ethical theories in response to contemporary issues and advancements in fields such as technology, medicine, and environmental ethics.

- **Applied Ethics:** Ethical theories inform practical applications in areas such as bioethics, business ethics, and political philosophy, guiding ethical decision-making and policy development.

Contemporary Relevance

1. Ethical Pluralism:

- **Diverse Perspectives:** Ethical pluralism acknowledges the coexistence of multiple ethical theories and perspectives within societies. Individuals and institutions may draw upon different ethical frameworks depending on context, values, and priorities.

- **Ethical Decision-Making:** Integrative approaches to ethics recognize the complexity of moral dilemmas and advocate for dialogue, reflection, and consensus-building across diverse perspectives.

2. Ethical Challenges:

- **Emerging Issues:** Contemporary ethical challenges, such as artificial intelligence ethics, global justice, and environmental sustainability, require nuanced ethical analyses and considerations from multiple ethical perspectives.

- **Cross-Cultural Dialogue:** Globalization fosters cross-cultural dialogue and ethical inquiry, challenging ethnocentric perspectives and promoting mutual understanding of diverse moral traditions and values.

The Influence of Religion and Philosophy

Throughout human history, religion and philosophy have served as foundational pillars in shaping moral and ethical frameworks that guide individual behavior and societal norms. These systems offer profound insights into the nature of morality, providing meaning, guidelines, and principles that shape human conduct. However, it is crucial to recognize that both religion and philosophy are products of human thought and culture, evolving over time and varying across different societies.

The Role of Religion

1. Moral Guidance and Sacred Texts:

- **Divine Command Ethics:** Many religious traditions, such as Judaism, Christianity, and Islam, emphasize moral principles derived from sacred texts and teachings attributed to divine authority. These texts, such as the Bible, Quran, or Vedas, provide ethical guidelines and narratives that shape believers' understanding of right and wrong.
- **Rituals and Ethics:** Religious rituals and practices reinforce ethical values, fostering communal identity and moral behavior through collective worship, prayer, and ethical teachings.

2. Moral Exemplars and Role Models:

- **Prophetic Figures:** Religious traditions often venerate moral exemplars and prophets who embody virtues and moral principles. These figures, such as Jesus Christ, Buddha, or Muhammad, serve as role models whose lives and teachings inspire ethical conduct and spiritual growth.

- **Community and Ethics:** Religious communities cultivate ethical norms through shared beliefs, communal practices, and moral teachings that promote virtues such as compassion, forgiveness, and justice.

The Influence of Philosophy

1. Ethical Theories and Philosophical Inquiry:

- **Ancient Greek Philosophy:** Philosophical traditions, from Socrates and Plato to Aristotle and the Stoics, have explored ethical questions through rational inquiry and critical reflection. Ethical theories like virtue ethics, consequentialism, and deontology provide frameworks for analyzing moral dilemmas and principles.

- **Human Reasoning:** Philosophical ethics emphasizes human reason and ethical principles that are accessible through rational discourse, empirical observation, and logical argumentation.

2. Secular Ethics and Modern Philosophy:

- **Enlightenment Thinkers:** During the Enlightenment era, philosophers such as Immanuel Kant and John Stuart Mill proposed secular ethical frameworks that prioritize reason, autonomy, and human rights. Kantian ethics, for example, emphasizes moral duties and the categorical imperative as universal principles of ethical behavior.

- **Utilitarianism:** Utilitarianism, advocated by Jeremy Bentham and John Stuart Mill, posits that actions are morally right if they maximize happiness or utility for the greatest number of people, emphasizing consequences and societal well-being.

Constructed Nature of Moral Systems

1. Cultural Context and Evolution:

- **Historical Development:** Moral and ethical systems evolve within cultural contexts, responding to social norms, historical events, and philosophical developments. They reflect human attempts to understand and justify moral principles that promote individual flourishing and communal well-being.

- **Pluralism and Diversity:** Globalization fosters ethical pluralism, where diverse

religious and philosophical traditions
coexist and interact, influencing ethical
discourse and cross-cultural
understanding.

2. Critical Reflection and Adaptation:

- **Ethical Challenges:** Contemporary
  ethical challenges, such as bioethics,
  environmental ethics, and social justice,
  require ongoing ethical reflection and
  adaptation. Religious and philosophical
  traditions engage with these issues,
  offering ethical insights and responses
  that address complex moral dilemmas.

- **Interfaith and Interphilosophical
  Dialogue:** Dialogue between religious
  and philosophical traditions promotes
  mutual understanding and ethical
  cooperation, fostering shared values of
  compassion, respect for human dignity,
  and social responsibility.

The Relativity of Morality

Morality, the principles that guide human conduct and ethical judgments, is not a fixed or universal standard. Instead, it varies significantly across cultures, societies, and historical epochs. This chapter explores the relativity of morality, highlighting how moral systems are constructed and shaped by diverse influences, including social norms, cultural values, and historical contexts.

Cultural Variability

1. Cultural Norms and Values:

- **Cultural Relativism:** Different cultures exhibit distinct moral norms and values that reflect their unique histories, traditions, and belief systems. What is considered morally acceptable or virtuous in one culture may be perceived differently in another.

- **Examples:** Practices such as polygamy, dietary restrictions, and rites of passage illustrate cultural variations in moral standards and ethical practices.

2. Ethical Pluralism:

- **Coexistence of Moral Systems:** Ethical pluralism acknowledges the coexistence of multiple moral frameworks within global society. Religious, philosophical, and secular traditions contribute to diverse ethical perspectives that inform individual and collective behavior.

- **Cross-Cultural Dialogue:** Interactions between cultures promote mutual understanding and challenge ethnocentric assumptions about morality, fostering respect for cultural diversity and ethical complexity.

Historical and Social Influences

1. Historical Context:

- **Evolution of Morality:** Morality evolves over time in response to historical events, social movements, and ideological shifts. Concepts of justice, human rights, and equality have evolved through historical struggles and moral debates.

- **Impact of Movements:** Abolitionism, civil rights movements, and women's suffrage illustrate how moral progress is shaped by social activism and ethical deliberation.

2. Social Constructionism:

- **Social and Institutional Factors:** Moral norms are reinforced and transmitted through social institutions such as family, education, religion, and law. These institutions uphold and perpetuate cultural values that shape individual moral development.

- **Power Dynamics:** Moral standards may reflect power dynamics and social hierarchies within societies, influencing who defines morality and whose voices are marginalized.

Challenges to Universality

1. Moral Relativism vs. Universalism:

- **Debates in Ethics:** Moral relativism contends that moral judgments are relative to cultural contexts and perspectives, challenging claims of universal moral truths. Universalism asserts the existence of objective moral principles applicable to all individuals, regardless of cultural background.

- **Ethical Dilemmas:** Globalization and multiculturalism raise ethical dilemmas that require navigating differences in moral perspectives, ethical practices, and legal frameworks across diverse contexts.

2. Moral Diversity and Dialogue:

- **Cross-Cultural Ethics:** Comparative ethics and intercultural dialogue explore ethical dilemmas and shared values across cultures, promoting ethical awareness, empathy, and mutual respect.

- **Ethical Globalization:** Ethical globalization emphasizes common ethical principles, such as human rights and environmental stewardship, that transcend cultural boundaries and promote global cooperation.

The relativity of morality underscores its
constructed nature, shaped by historical, social,
and cultural influences. Recognizing the
diversity of moral systems encourages humility,
empathy, and critical reflection on ethical
principles. By engaging in cross-cultural
dialogue and ethical inquiry, we deepen our
understanding of morality's complexity and
foster a more inclusive and ethical global
community that respects cultural diversity while
striving for common ethical ideals. This chapter
invites readers to consider the dynamic nature
of morality and its profound impact on human
interactions and societal norms.

Chapter 8: Technology - Creating New Realities

The Evolution of Technology

Technology extends human capabilities and shapes our interaction with the world. From the wheel to the internet, technological advancements have transformed human societies, creating new constructs and realities.

Early Technological Advancements

1. Ancient Innovations:

- **Early Tools and Agriculture:** Human history began with the development of simple tools and techniques for hunting, gathering, and agriculture. These innovations laid the foundation for settled societies and the division of labor.
- **Writing and Communication:** The invention of writing systems, such as

cuneiform in Mesopotamia and hieroglyphs in Egypt, enabled the recording of knowledge, laws, and cultural practices, fostering intellectual and cultural development.

2. Mechanical Inventions:

- **Ancient Engineering:** Civilizations like ancient Egypt and Rome pioneered engineering feats such as aqueducts, roads, and architectural marvels like the pyramids and colosseums. These constructions demonstrated early mastery of materials and structural design.

Industrial Revolution and Technological Revolution

1. Industrial Revolution (18th-19th centuries):

- **Mechanization and Manufacturing:** The Industrial Revolution marked a shift from agrarian economies to industrialized societies. Innovations such as steam engines, textile machinery, and iron production mechanized production and transportation, driving urbanization and economic growth.

- **Impact on Society:** Industrialization spurred social changes, including urban migration, labor reforms, and new social classes, while also fueling scientific inquiry and technological innovation.

2. Technological Revolution (20th century-present):

- **Electronics and Information Age:** The 20th century saw rapid advancements in electronics, telecommunications, and computing. The invention of the transistor, development of integrated circuits, and the rise of computers transformed communication, computation, and data storage capabilities.

- **Digital Age:** The advent of the internet and digital technologies revolutionized global connectivity, commerce, and information sharing, reshaping industries, education, and social interactions.

Contemporary Technological Trends

1. Information Technology:

- **Internet and Connectivity:** The proliferation of smartphones, broadband internet, and wireless technologies has connected billions worldwide, enabling instant communication, access to information, and digital commerce.

- **Artificial Intelligence and Big Data:** Advances in AI, machine learning, and big data analytics are driving innovations in automation, personalized services, healthcare diagnostics, and autonomous systems.

2. Biotechnology and Healthcare:

- **Genomics and Medical Advances:** Biotechnological breakthroughs, such as CRISPR gene editing and personalized medicine, are revolutionizing healthcare by offering tailored treatments, disease prevention, and genetic research capabilities.

- **Medical Devices:** Innovations in medical devices, robotics, and prosthetics are enhancing patient care, surgical precision, and rehabilitation outcomes.

Societal Impacts and Challenges

1. Globalization and Economy:

- **Digital Economy:** The rise of e-commerce, digital platforms, and fintech innovations is reshaping global trade, consumer behavior, and financial services, fostering economic growth and entrepreneurship.

- **Job Displacement and Skills Gap:** Automation and AI advancements raise concerns about job displacement and the need for reskilling and education in digital literacy and technical skills.

2. Ethical and Environmental Considerations:

- **Ethical Challenges:** Technological advancements pose ethical dilemmas regarding privacy, data security, algorithmic biases, and the ethical use of AI and biotechnology.

- **Environmental Sustainability:** Sustainable technologies, renewable energy solutions, and efforts to mitigate technological carbon footprints are critical in addressing environmental impacts and climate change.

Future Directions and Innovation

1. Emerging Technologies:

- **Space Exploration:** Advances in aerospace technology, private space ventures, and exploration missions aim to expand human presence beyond Earth and unlock the mysteries of the universe.
- **Emerging Fields:** Nanotechnology, quantum computing, 3D printing, and renewable energy innovations hold promise for addressing global challenges and transforming industries.

2. Ethical Innovation:

- **Responsible Development:** As technology continues to evolve, ethical frameworks and regulations must evolve alongside to ensure responsible innovation, protect societal values, and promote equitable access to benefits.

Virtual Reality and Simulations

Virtual reality (VR) and simulations represent a paradigm shift in human experience, offering immersive environments that challenge our perception of reality and expand the possibilities of human interaction and exploration. This chapter explores how VR and simulations, as human-made constructs, create new realities and redefine our understanding of existence and interaction.

Immersive Experiences

1. Virtual Environments:

- **Technological Advancements:** VR technologies use computer-generated environments to simulate sensory experiences, including sight, sound, and touch, creating a sense of presence in virtual worlds.

- **Applications:** From entertainment and gaming to education, healthcare, and training, VR enhances immersive learning, therapy, and skill development in controlled and interactive environments.

2. Simulation Technologies:

- **Modeling Complex Systems:** Simulations replicate real-world processes, behaviors, and interactions to study and predict outcomes in fields such as engineering, climate science, economics, and medicine.
- **Training and Testing:** Simulations enable training simulations for pilots, surgeons, and emergency responders, offering safe and repeatable scenarios to refine skills and decision-making.

Blurring Real and Virtual Realities

1. Psychological Impact:

- **Sense of Presence:** VR's immersive nature can evoke emotional responses and a sense of presence in virtual environments, blurring distinctions between real and simulated experiences.
- **Ethical Considerations:** Ethical dilemmas arise regarding the psychological effects of prolonged immersion in virtual worlds and the potential for addiction or dissociation from reality.

2. Philosophical Reflections:

- **Nature of Reality:** VR challenges philosophical inquiries into the nature of reality and consciousness, raising questions about the authenticity of experiences and perceptions within virtual environments.

- **Identity and Self:** Virtual identities and avatars in VR spaces explore notions of digital identity, embodiment, and the boundaries between physical and virtual selves.

Constructed Realities

1. Human-Made Constructs:

- **Designing Virtual Worlds:** VR developers design virtual worlds with rules, environments, and interactions that shape user experiences and narratives, illustrating how human creativity constructs alternative realities.

- **Cultural and Social Dynamics:** Virtual communities and social interactions within VR platforms reflect cultural norms, social hierarchies, and collaborative behaviors, influencing user experiences and community dynamics.

2. Artistic Expression and Creativity:

- **Digital Art and Design:** VR serves as a canvas for artistic expression and creativity, enabling immersive storytelling, interactive narratives, and spatial experiences that transcend traditional mediums.

- **Virtual Museums and Galleries:** VR applications in cultural institutions offer virtual tours, exhibitions, and archival experiences, democratizing access to art, history, and heritage globally.

Future Directions and Ethical Challenges

1. Technological Innovations:

- **Augmented Reality (AR):** AR integrates digital information into the user's real-world environment, enhancing perception and interaction with physical surroundings through wearable devices and smartphones.

- **Mixed Reality (MR):** MR blends virtual and physical realities, enabling interactive experiences that merge virtual objects with real-world environments, revolutionizing industries like architecture, retail, and entertainment.

2. Ethical Considerations:

- **Privacy and Surveillance:** VR raises concerns about data privacy, surveillance, and the ethical use of user data in virtual environments.
- **Ethical Design:** Responsible VR development requires ethical guidelines to address issues of inclusivity, accessibility, cultural sensitivity, and user well-being in virtual experiences.

The Future of Artificial Intelligence

Artificial Intelligence (AI) stands at the forefront of technological innovation, transforming industries, enhancing productivity, and reshaping human interaction with machines. This chapter explores how AI represents a frontier in constructing new realities, challenging our understanding of consciousness, intelligence, and the very nature of reality itself.

The Rise of Artificial Intelligence

1. Evolution of AI:

- **From Concept to Reality:** AI has evolved from theoretical concepts in the mid-20th century to practical applications today. Early developments in machine learning, neural networks, and algorithms laid the foundation for AI's exponential growth.

- **Technological Advancements:** Advances in computing power, big data analytics, and cloud computing have fueled AI's capabilities in processing vast amounts of data, learning from patterns, and making autonomous decisions.

2. Applications Across Industries:

- **Automation and Efficiency:** AI technologies automate routine tasks, optimize processes, and enhance efficiency in industries such as manufacturing, logistics, finance, and healthcare.
- **Personalization and User Experience:** AI powers recommendation systems, virtual assistants, and personalized

content delivery, improving user experience and engagement in digital platforms.

Challenges to Understanding AI

1. Consciousness and Intelligence:

- **Artificial General Intelligence (AGI):** The quest for AGI aims to develop AI systems capable of human-like reasoning, learning, and problem-solving across diverse domains.

- **Ethical and Philosophical Implications:** AI's potential to exhibit autonomous decision-making and creativity raises questions about consciousness, moral agency, and the implications for human identity and society.

2. Ethical Considerations:

- **Bias and Fairness:** AI algorithms can perpetuate biases present in training data, impacting decisions in hiring, finance, law enforcement, and healthcare.

- **Transparency and Accountability:** The opacity of AI decision-making processes raises concerns about accountability, ethical governance, and the need for transparency in algorithmic outcomes.

Constructing New Realities with AI

1. Simulation and Prediction:

- **Predictive Analytics:** AI models predict outcomes, trends, and behaviors based on historical data, influencing decision-making in business, policy-making, and strategic planning.

- **Climate Modeling:** AI contributes to climate science by modeling complex environmental systems, informing climate change mitigation strategies and sustainability efforts.

2. Creative Expression and Innovation:

- **AI in Art and Design:** AI generates art, music, and literature, blurring distinctions between human creativity and machine-generated content. Collaborations between artists and AI explore new forms of artistic expression and creativity.

- **Innovation and Discovery:** AI accelerates scientific discovery by analyzing genomic data, simulating molecular structures, and identifying new materials with potential applications in medicine, materials science, and renewable energy.

Future Directions and Societal Impact

1. Human-AI Collaboration:

- **Augmented Intelligence:** Human-AI partnerships leverage AI's computational capabilities and human expertise to enhance decision-making, creativity, and problem-solving in diverse fields.

- **Skills and Workforce:** AI-driven automation and augmentation reshape the workforce, requiring reskilling and upskilling initiatives to adapt to evolving job roles and technological advancements.

2. Ethical Governance and Regulation:

- **Ethical Frameworks:** Developing ethical guidelines and regulations ensures responsible AI development, addressing issues of privacy, bias, safety, and societal impact.

- **Global Collaboration:** International cooperation fosters standards for AI governance, promotes transparency in AI deployment, and mitigates risks associated with AI's rapid evolution and adoption.

Artificial Intelligence represents a transformative frontier in constructing new realities, challenging traditional notions of consciousness, intelligence, and the fabric of reality itself. As AI technologies continue to advance, it is imperative to navigate ethical considerations, foster human-AI collaboration, and ensure responsible deployment to harness AI's potential for innovation, societal benefit, and ethical integrity. Understanding AI's impact on constructing new realities requires interdisciplinary dialogue, ethical reflection, and proactive governance to shape a future where AI enhances human capabilities while respecting fundamental values and principles.

Technology's Role in Shaping Perception

Technology has become integral to how we perceive and interact with the world, transforming our experiences and realities through digital constructs that influence communication, information consumption, and personal identity. This chapter explores the profound impact of technology on shaping perception, from social media dynamics to augmented reality, highlighting how digital innovations redefine human interaction and understanding.

Digital Constructs and Perception

1. Social Media Dynamics:

- **Digital Identities:** Social media platforms create virtual personas based

on curated content, interactions, and
social connections, shaping how
individuals perceive themselves and
others.

- **Filtering Reality:** Algorithms prioritize
content based on user preferences and
engagement, filtering information and
reinforcing echo chambers that influence
worldview and perception of societal
issues.

2. Information Consumption:

- **Personalized Content:** AI-driven
recommendation systems tailor news,
entertainment, and advertisements to
individual preferences, influencing
perception of current events and
consumer choices.

- **Confirmation Bias:** Access to diverse
viewpoints competes with echo
chambers, challenging users to critically

evaluate information and perspectives
encountered online.

Augmented Reality and Virtual Environments

1. Immersive Experiences:

- **Augmented Reality (AR):** AR overlays
  digital information onto the physical
  world through mobile devices and
  wearable technology, enhancing
  real-world interactions and spatial
  perception.
- **Virtual Reality (VR):** VR transports
  users to simulated environments, offering
  immersive experiences that alter sensory
  perception and spatial awareness,
  blurring distinctions between physical
  and digital realities.

2. Impact on Education and Training:

- **Interactive Learning:** AR and VR applications in education create interactive simulations and virtual classrooms that enhance engagement, spatial reasoning, and experiential learning.
- **Professional Development:** VR training programs simulate real-world scenarios for medical, military, and industrial training, improving skill acquisition and performance in high-risk environments.

Ethical and Societal Implications

1. Privacy and Surveillance:

- **Data Privacy Concerns:** Technologies like facial recognition, location tracking, and biometric data collection raise ethical concerns about surveillance, personal privacy, and data security.

- **Regulatory Challenges:** Balancing innovation with regulatory frameworks ensures responsible use of technology while safeguarding individual rights and freedoms in digital spaces.

2. Psychological and Social Well-being:

- **Digital Addiction:** Prolonged engagement with digital devices and virtual environments contributes to screen addiction and impacts mental health, emphasizing the need for digital balance and mindful technology use.

- **Social Relationships:** Online interactions and virtual communities influence social norms, communication styles, and interpersonal relationships, redefining social dynamics and community engagement.

Future Directions and Human-Centered Design

1. Human-Centered Technology:

- **User Experience Design:** Prioritizing user needs and ethical considerations in technology development promotes inclusive, accessible, and user-friendly digital experiences.

- **Designing for Empathy:** Emphasizing empathy and human connection in technology design fosters meaningful interactions, digital literacy, and responsible technology use.

2. Ethical Leadership and Innovation:

- **Corporate Responsibility:** Tech companies play a pivotal role in ethical leadership, transparency, and accountability in AI, data privacy, and digital innovation.

- **Global Collaboration:** International cooperation and standards promote ethical guidelines, regulatory frameworks, and technological governance that prioritize human rights and societal well-being.

Technology's role in shaping perception underscores its transformative impact on how individuals, communities, and societies perceive and interact with the world. From social media dynamics to augmented reality experiences, digital constructs redefine human experiences, influence societal norms, and challenge ethical considerations. Embracing human-centered design principles, fostering digital literacy, and navigating ethical complexities are essential in harnessing technology's potential to enrich human perception, foster innovation, and promote inclusive, ethical digital futures. Understanding technology's influence on perception requires critical reflection, interdisciplinary collaboration, and proactive engagement to shape a future where technology enhances human experiences while preserving fundamental values and ethical principles.

Chapter 9: The Psychology of Belief

How Beliefs Are Formed

Beliefs are the bedrock of our perception of reality, influencing our actions, decisions, and interactions with the world. This chapter explores the complex processes through which beliefs are formed, including personal experiences, cultural influences, and cognitive mechanisms. Understanding how beliefs are constructed sheds light on their powerful role in shaping human behavior and societal dynamics.

Personal Experiences

1. Direct Experiences:

- **Sensory Input:** Our beliefs often stem
  from direct sensory experiences. What
  we see, hear, touch, taste, and smell
  provide the raw data that our brains
  interpret to form beliefs about the world
  around us.

- **Life Events:** Significant life events, such
  as personal achievements, failures,
  trauma, or relationships, play a crucial
  role in shaping our beliefs. These
  experiences create lasting impressions
  that influence how we perceive and react
  to future situations.

2. Observational Learning:

- **Social Learning Theory:** Observing the
  behaviors and outcomes of others can
  shape our beliefs. This process, known as

observational learning, is a fundamental mechanism through which beliefs about social norms, risks, and rewards are formed.

- **Role Models:** Influential figures in our lives, such as parents, teachers, and peers, serve as role models whose beliefs and behaviors we often emulate. Their influence is particularly strong during childhood and adolescence.

Cultural Influences

1. Societal Norms and Values:

- **Cultural Frameworks:** The culture we are born into provides a framework of norms, values, and traditions that shape our beliefs. These cultural constructs offer guidelines on what is considered right or wrong, acceptable or unacceptable.

- **Socialization:** From a young age, individuals are socialized into their culture through family, education, religion, and media. This process instills shared beliefs that reinforce group identity and cohesion.

2. Language and Communication:

- **Linguistic Relativity:** The language we speak influences how we perceive and conceptualize the world, a concept known as linguistic relativity or the Sapir-Whorf hypothesis. Language shapes our thoughts and, consequently, our beliefs.

- **Narratives and Stories:** Stories and narratives, whether told through literature, folklore, or media, play a significant role in shaping beliefs. They provide context and meaning, helping individuals make sense of their experiences and the world.

Cognitive Processes

1. Cognitive Biases:

- **Confirmation Bias:** One of the most well-known cognitive biases, confirmation bias, leads individuals to seek out, interpret, and remember information that confirms their preexisting beliefs while ignoring or discounting contradictory evidence.
- **Anchoring and Framing:** The way information is presented (framing) and the initial information received (anchoring) can heavily influence belief formation. These cognitive biases affect decision-making and judgment.

2. Heuristics and Mental Shortcuts:

- **Simplifying Complexity:** Heuristics are mental shortcuts that help individuals make decisions and form beliefs quickly. While useful, they can also lead to systematic errors and biases.
- **Pattern Recognition:** Humans have an innate tendency to recognize patterns and make connections, sometimes seeing correlations where none exist. This pattern recognition can lead to the formation of superstitious or unfounded beliefs.

The Impact of Beliefs on Perception and Behavior

1. Shaping Reality:

- **Perceptual Filters:** Beliefs act as filters through which we interpret sensory information. They influence what we notice, how we interpret events, and what we remember, effectively shaping our perceived reality.

- **Self-Fulfilling Prophecies:** Beliefs can lead to self-fulfilling prophecies, where an individual's expectations influence their behavior in a way that causes those expectations to come true.

2. Guiding Actions and Decisions:

- **Decision-Making:** Beliefs play a critical role in decision-making processes, guiding choices in personal, professional, and societal contexts. They provide a framework for evaluating options and predicting outcomes.
- **Behavioral Consistency:** Individuals tend to act in ways that are consistent with their beliefs. This consistency

reinforces the stability of belief systems over time.

Cognitive Biases and Illusions

Human perception and understanding of reality are heavily influenced by cognitive biases, which are systematic patterns of deviation from norm or rationality in judgment. Cognitive biases such as confirmation bias and anchoring affect how we interpret information and form beliefs, demonstrating the constructed nature of our perceptions and understanding of reality. This chapter delves into these biases and their implications for our cognitive processes and worldviews.

The Nature of Cognitive Biases

1. Definition and Overview:

- **Systematic Deviations:** Cognitive biases are tendencies to think in certain ways that can lead to systematic deviations from a standard of rationality or good judgment. These biases are often a result of the brain's attempt to simplify information processing.

- **Evolutionary Roots:** Many cognitive biases have evolutionary roots, having developed as mechanisms to help our ancestors make quick decisions in complex and uncertain environments.

Key Cognitive Biases

1. Confirmation Bias:

- **Selective Attention:** Confirmation bias is the tendency to search for, interpret, and remember information in a way that confirms one's preexisting beliefs or hypotheses. This bias leads individuals to give more weight to evidence that

supports their beliefs and to disregard or
downplay evidence that contradicts them.

- **Impact on Belief Systems:**
  Confirmation bias reinforces existing
  beliefs and makes it difficult for
  individuals to change their views, even in
  the face of contradictory evidence. This
  bias can perpetuate misinformation and
  hinder critical thinking.

2. Anchoring Bias:

- **Initial Information:** Anchoring bias
  occurs when individuals rely too heavily
  on the first piece of information they
  receive (the "anchor") when making
  decisions. Subsequent judgments are
  influenced by this initial anchor,
  regardless of its relevance or accuracy.
- **Decision-Making:** Anchoring can affect
  a wide range of decisions, from everyday
  choices to significant financial and legal

judgments. It highlights how initial impressions can disproportionately shape subsequent thinking and behavior.

The Constructed Nature of Perception

1. Interpretative Filters:

- **Filtering Information:** Cognitive biases act as filters through which we interpret and understand information. These biases shape our perception by influencing what we pay attention to and how we interpret data.

- **Constructed Reality:** The interplay of various cognitive biases constructs our subjective reality, shaping how we perceive and interact with the world. This constructed nature underscores the idea that our understanding of reality is not objective but influenced by inherent mental shortcuts and biases.

2. Social and Cultural Reinforcement:

- **Echo Chambers:** Social and cultural environments can reinforce cognitive biases. For instance, online echo chambers and filter bubbles, created by algorithms that prioritize content aligned with users' beliefs, can intensify confirmation bias.
- **Cultural Norms:** Cultural norms and values shape the collective biases of a group, influencing shared perceptions and beliefs. These collective biases can reinforce societal structures and ideologies.

Implications for Understanding Reality

1. Limits of Objectivity:

- **Subjective Interpretation:** The influence of cognitive biases highlights the limitations of objectivity in human perception. Our understanding of reality is shaped by subjective interpretations and mental constructs.

- **Critical Thinking:** Recognizing cognitive biases is crucial for fostering critical thinking and self-awareness. It encourages individuals to question their assumptions and seek diverse perspectives to mitigate the effects of these biases.

2. Decision-Making and Judgment:

- **Improving Decisions:** Understanding cognitive biases can improve decision-making by promoting awareness of potential pitfalls and encouraging strategies to counteract biased thinking.
- **Policy and Education:** Integrating knowledge of cognitive biases into

education and policymaking can enhance
critical thinking skills and promote
informed decision-making at individual
and societal levels.

Strategies to Mitigate Cognitive Biases

1. Awareness and Education:

- **Cognitive Training:** Educating
  individuals about cognitive biases and
  their effects can raise awareness and
  promote strategies to counteract biased
  thinking.
- **Critical Reflection:** Encouraging critical
  reflection on one's beliefs and
  decision-making processes can help
  identify and address cognitive biases.

2. Diverse Perspectives:

- **Exposure to Diversity:** Seeking out
  diverse perspectives and engaging with

viewpoints different from one's own can challenge confirmation bias and broaden understanding.

- **Collaborative Decision-Making:** Collaborative decision-making processes that incorporate diverse opinions can mitigate the influence of individual biases and lead to more balanced outcomes.

The Power of the Mind in Shaping Reality

The human mind wields significant power in shaping our perception of reality. Phenomena such as the placebo effect and mass hysteria illustrate how beliefs and expectations can influence both physical and psychological experiences. This chapter explores the profound impact of the mind on our understanding of the world and our experiences within it, highlighting the intricate relationship between cognition, belief, and reality.

The Placebo Effect

1. Understanding the Placebo Effect:

- **Definition:** The placebo effect occurs when a person experiences a real improvement in their condition after receiving a treatment that has no therapeutic effect. This improvement is driven by the individual's belief in the efficacy of the treatment.
- **Mechanisms:** The placebo effect is thought to work through several mechanisms, including the release of endorphins, changes in brain activity, and psychological factors such as expectation and conditioning.

2. Evidence and Examples:

- **Medical Studies:** Numerous clinical trials have demonstrated the placebo effect in various conditions, from pain management to depression. Placebos have been shown to produce measurable physiological changes, such as altered heart rates and hormone levels.

- **Impact on Treatment:** The placebo effect underscores the importance of patient perception and belief in the effectiveness of medical treatments. It highlights the mind's ability to influence health outcomes through psychological and physiological pathways.

Mass Hysteria

1. Defining Mass Hysteria:

- **Description:** Mass hysteria, or collective hysteria, refers to the rapid spread of illness signs and symptoms within a cohesive group, where there is no identifiable organic cause. It often results from psychological stress and the power of suggestion.

- **Historical Examples:** Historical instances of mass hysteria include the Salem witch trials, the dancing plague of 1518, and various cases of epidemic hysteria in schools and workplaces.

2. Psychological Mechanisms:

- **Social Contagion:** Mass hysteria is driven by social contagion, where emotions and behaviors spread rapidly through a group. Fear, anxiety, and suggestibility play crucial roles in the development and propagation of symptoms.
- **Group Dynamics:** Factors such as group cohesion, social reinforcement, and the presence of authority figures can amplify the spread of mass hysteria. Individuals' perceptions are influenced by the

reactions and beliefs of those around them.

The Influence of Beliefs and Expectations

1. Cognitive and Emotional Impact:

- **Expectation and Perception:** Beliefs and expectations shape how we perceive and interpret sensory information. Positive expectations can lead to improved outcomes, while negative expectations can exacerbate symptoms or create new ones.
- **Emotion and Cognition:** Emotions, such as fear and hope, significantly influence our cognitive processes. They affect attention, memory, and decision-making, further shaping our perception of reality.

2. Real-World Implications:

- **Healthcare:** The placebo effect demonstrates the importance of patient beliefs in healthcare. Enhancing patient trust and positive expectations can improve treatment outcomes, even when the treatment itself is inert.

- **Social Behavior:** Understanding mass hysteria provides insight into how social dynamics and psychological factors can lead to widespread irrational behavior. This knowledge is crucial for managing public health responses and preventing panic during crises.

The Constructed Nature of Reality

1. Perception as Construction:

- **Sensory Inputs:** Our brains construct a coherent reality from sensory inputs, filling in gaps and interpreting

ambiguous information based on past
experiences and beliefs.

- **Subjective Reality:** The constructed
nature of perception means that reality is
subjective and can vary significantly
between individuals based on their
cognitive and emotional states.

2. The Role of Cognitive Biases:

- **Confirmation Bias:** Cognitive biases,
such as confirmation bias, reinforce
existing beliefs and expectations, further
shaping perception and interpretation of
events.
- **Anchoring Effect:** Initial information or
experiences (anchors) heavily influence
subsequent perceptions and judgments,
highlighting the mind's role in
constructing reality.

Harnessing the Power of the Mind

1. Positive Applications:

- **Therapeutic Interventions:** Techniques such as cognitive-behavioral therapy (CBT) leverage the power of beliefs and expectations to treat mental health conditions. By altering negative thought patterns, CBT can lead to significant improvements in well-being.
- **Mindfulness and Meditation:** Practices like mindfulness and meditation harness the mind's power to reduce stress, enhance focus, and improve emotional regulation. These practices demonstrate the beneficial effects of positive cognitive and emotional states.

2. Ethical Considerations:

- **Placebo Use:** The ethical use of placebos in clinical practice raises questions about patient consent and the balance between belief-driven benefits and deception.
- **Public Information:** Managing mass hysteria and the spread of misinformation requires ethical considerations in communication and public health strategies to prevent panic and promote rational decision-making.

Chapter 10: Deconstructing Reality - What Remains?

Philosophical Deconstruction

Philosophical deconstruction involves analyzing and breaking down constructs to understand their foundations and implications. This process, rooted in the works of thinkers such as Jacques Derrida, seeks to uncover the underlying assumptions and contradictions within established concepts and systems. By dissecting these constructs, deconstruction reveals the arbitrary and constructed nature of many aspects of reality that we often take for granted.

One of the central tenets of deconstruction is the idea that language shapes our understanding of reality. Words and symbols are not neutral conveyors of meaning but are loaded with cultural, historical, and ideological baggage.

Through the process of deconstruction, we can uncover the multiple, often conflicting meanings that words and concepts carry. This reveals how our perceptions of truth and reality are constructed through language, rather than being direct reflections of an objective world.

Deconstruction also challenges the notion of binary oppositions, which are pairs of opposing concepts that are often seen as natural and mutually exclusive. Examples of such binaries include good/evil, male/female, and nature/culture. Deconstruction shows that these oppositions are not inherent in the world but are constructed through social and cultural practices. By breaking down these binaries, deconstruction exposes the power dynamics and hierarchies that they support.

Furthermore, deconstruction highlights the instability and fluidity of meaning. It suggests that meanings are not fixed or stable but are constantly shifting and evolving. This challenges the idea that any concept or text has a single, definitive interpretation. Instead, deconstruction encourages us to see meaning as a dynamic and contested process, open to multiple interpretations and reinterpretations.

The implications of deconstruction extend beyond philosophy and into various fields, including literature, politics, and social sciences. In literature, deconstruction has been used to analyze texts, revealing the hidden assumptions and contradictions within them. In politics, deconstruction can be used to critique ideologies and power structures, exposing how they are constructed and maintained through discourse. In social sciences, deconstruction helps to uncover the ways in which social categories and identities are constructed and contested.

Deconstruction also invites us to question the foundations of knowledge and truth. It challenges the idea that knowledge is a neutral and objective pursuit, showing instead how it is shaped by cultural and historical contexts. This has significant implications for fields such as science, where deconstruction can be used to critique the assumptions and biases that underpin scientific knowledge and practices.

Moreover, deconstruction encourages a more critical and reflective approach to our own beliefs and assumptions. By revealing the constructed nature of reality, deconstruction prompts us to examine the foundations of our own thinking and to question the certainties that we often take for granted. This can lead to a more open-minded and flexible approach to knowledge and understanding.

However, deconstruction is not without its critics. Some argue that it leads to relativism, where all interpretations are seen as equally valid, undermining the possibility of objective truth or moral certainty. Others suggest that deconstruction can be overly negative and destructive, focusing on tearing down constructs without offering alternatives. Despite these criticisms, deconstruction remains a powerful tool for analyzing and understanding the complexities of reality.

Philosophical deconstruction involves breaking down constructs to understand their foundations and implications. This process reveals the arbitrary and constructed nature of many aspects of reality, challenging our assumptions and prompting us to think more critically and reflectively. By uncovering the ways in which language, binary oppositions, and power dynamics shape our perceptions of truth and reality, deconstruction opens up new possibilities for understanding and engaging with the world.

The Concept of Nothingness

Exploring the concept of nothingness challenges our understanding of existence and reality, pushing the boundaries of philosophical inquiry and human thought. Nothingness, or the absence of being, is a profound and often unsettling concept that forces us to confront the limitations of our perception and the foundations of our existence. By examining philosophical perspectives such as existentialism and nihilism, we can gain deeper insights into the implications of deconstructing reality and understanding the void.

The Nature of Nothingness

Nothingness is not merely the absence of physical objects or entities; it represents the absence of anything at all, including space, time, and existence itself. This concept transcends the material world and delves into the metaphysical realm, questioning the very fabric of reality. Philosophers have long grappled with the idea of nothingness, attempting to define and understand its role in the context of existence. Nothingness challenges our intuitive notions of reality, forcing us to reconsider what it means to exist and to be.

Existentialism and the Void

Existentialist philosophers such as Jean-Paul Sartre and Martin Heidegger have explored the concept of nothingness extensively. In existentialism, nothingness is closely linked to the human condition and the search for meaning. Sartre, for example, argued that nothingness is an inherent part of human existence. He posited that we are constantly confronted with the void—the absence of inherent meaning in life—which compels us to create our own purpose and values. This confrontation with nothingness is both a source of anxiety and a catalyst for authentic living.

Nihilism and the Denial of Meaning

Nihilism, on the other hand, takes a more radical stance on nothingness. Nihilistic thinkers, such as Friedrich Nietzsche, assert that life is inherently meaningless and that any attempt to find purpose is ultimately futile. For nihilists, nothingness is not just an abstract concept but a fundamental truth about the nature of existence. This perspective denies the possibility of inherent meaning or value in the universe, leading to a profound sense of disillusionment and despair. However, Nietzsche also saw the potential for liberation in embracing nothingness, as it frees individuals from the constraints of imposed values and beliefs.

Deconstructing Reality

Deconstructing reality through the lens of nothingness involves questioning the very foundations of our understanding of the world. This process reveals the arbitrary and constructed nature of many aspects of reality that we often take for granted. By peeling back the layers of meaning and existence, we expose the underlying void that lies at the heart of our perceptions and beliefs. This deconstruction can be both disorienting and enlightening, as it forces us to confront the limits of our knowledge and the fragility of our constructed realities.

The Psychological Impact of Nothingness

The concept of nothingness has significant psychological implications. Confronting the void can evoke feelings of existential dread and anxiety, as it challenges our sense of purpose and stability. However, it can also lead to a deeper understanding of ourselves and our place in the universe. By acknowledging the nothingness that underpins existence, we can cultivate a more nuanced and authentic approach to life, grounded in the acceptance of uncertainty and the recognition of our freedom to create meaning.

The Role of Nothingness in Art and Literature

Nothingness has been a recurring theme in art and literature, reflecting humanity's ongoing struggle to grapple with the void. From the existential musings of Kafka and Camus to the abstract expressions of modern art, creators have used their work to explore the implications of nothingness and its impact on the human condition. Through these creative expressions, we gain a deeper appreciation of the ways in which nothingness shapes our perceptions, emotions, and experiences.

Nothingness and Spirituality

In many spiritual traditions, nothingness is a central concept that represents the ultimate reality or the ground of being. For example, in Buddhism, the idea of emptiness (Śūnyatā) is a fundamental teaching that emphasizes the impermanent and interdependent nature of all phenomena. Embracing the concept of nothingness in a spiritual context can lead to profound insights and a deeper sense of peace and enlightenment. It encourages the dissolution of the ego and the transcendence of dualistic thinking, fostering a more holistic understanding of existence.

The Scientific Perspective on Nothingness

From a scientific standpoint, nothingness also presents intriguing questions. In cosmology, the concept of the vacuum state—space devoid of matter—raises questions about the nature of the universe and the origins of existence. Quantum mechanics further complicates our understanding of nothingness, suggesting that even in a vacuum, particles and antiparticles constantly flicker in and out of existence. These scientific explorations highlight the complexity and mystery of nothingness, challenging our conventional notions of reality.

Embracing Nothingness in Everyday Life

Integrating the concept of nothingness into our everyday lives involves embracing uncertainty and recognizing the constructed nature of our beliefs and values. By doing so, we can cultivate a more flexible and adaptive mindset, better equipped to navigate the complexities of existence. Embracing nothingness also encourages us to let go of rigid attachments and to approach life with a sense of openness and curiosity. This mindset fosters resilience and creativity, allowing us to find meaning and purpose even in the face of the void.

Embracing the Absurd

Embracing the absurd involves accepting the inherent contradictions and uncertainties of life. This perspective, rooted in existentialist thought and popularized by philosophers like Albert Camus, encourages individuals to find meaning and purpose even in the face of a constructed and often illusory reality. Joshua Douthett's philosophy of flowism complements this approach, advocating for a harmonious alignment with the natural flow of life. By integrating the principles of embracing the absurd and flowism, we can navigate the complexities of existence with resilience and grace.

The Absurd Condition

The concept of the absurd arises from the recognition that life is filled with contradictions and uncertainties. According to Camus, the absurd is the conflict between our desire for meaning and the indifferent universe that offers none. This realization can lead to a sense of existential disorientation, as we grapple with the futility of seeking absolute truths in a world that defies simple explanations. Embracing the absurd means acknowledging this conflict without succumbing to despair.

Finding Meaning Amid Contradictions

While the absurd condition may seem daunting, it also presents an opportunity for profound personal growth. By accepting the inherent contradictions and uncertainties of life, we can liberate ourselves from the need for definitive answers. Instead of searching for an elusive ultimate meaning, we can create our own purpose and values. This self-created meaning becomes a source of strength and resilience, allowing us to navigate life's challenges with greater clarity and intention.

Flowism and Harmonious Living

Flowism, as popularized by Joshua Douthett, complements the philosophy of embracing the absurd by promoting a harmonious alignment with the natural flow of life. Flowism emphasizes living in tune with the rhythms and patterns of existence, rather than resisting or attempting to control them. By going with the flow, individuals can experience a sense of ease and connection, even in the face of uncertainty and contradiction.

The concepts of embracing the absurd and adopting a flow-oriented mindset not only provide a framework for living harmoniously but also offer profound insights into how we navigate the illusion of reality. By accepting the inherent contradictions and uncertainties of existence and aligning ourselves with the natural flow of life, we can transcend the superficial layers of reality and tap into a deeper, more authentic experience. This chapter explores how these philosophies help us understand and move through the constructed nature of our perceptions and beliefs, ultimately guiding us toward a more fulfilling and enlightened existence.

The Illusion of Reality

Reality, as we perceive it, is a construct shaped by our sensory experiences, cognitive processes, and cultural influences. Our brains filter and interpret the vast array of information we encounter, creating a coherent narrative that allows us to function in the world. However, this narrative is not an objective representation of reality but a subjective interpretation influenced by our beliefs, biases, and societal norms. By recognizing the constructed nature of reality, we can begin to question our assumptions and open ourselves to new ways of seeing and understanding the world.

Embracing the Absurd to See Beyond Constructs

Embracing the absurd involves accepting the inherent contradictions and uncertainties of life. This perspective encourages us to question the fixed meanings and truths we often take for granted. By acknowledging the absurdity of our search for definitive answers in an indifferent universe, we can free ourselves from the constraints of rigid thinking. This liberation allows us to see beyond the constructs of reality and appreciate the fluid and dynamic nature of existence. It encourages us to create our own meaning and values, rather than relying on external validation.

Flowism and the Fluidity of Reality

Flowism, as advocated by Joshua Douthett, emphasizes living in harmony with the natural rhythms and patterns of life. This philosophy teaches us to align ourselves with the flow of events rather than resisting or attempting to control them. By adopting a flow-oriented mindset, we can navigate the ever-changing landscape of reality with greater ease and adaptability. This approach helps us recognize the transient nature of our experiences and the constructed nature of our perceptions. It encourages us to be present in the moment and to respond to life's challenges with flexibility and resilience.

Practical Implications for Daily Life

Integrating the principles of embracing the absurd and flowism into our daily lives has practical implications for how we navigate the illusion of reality. By accepting uncertainty and aligning with the flow, we can reduce stress and anxiety, enhance our creativity, and build more meaningful relationships. This approach fosters a sense of inner peace and fulfillment, as we let go of the need for rigid control and embrace the dynamic nature of existence. It empowers us to create our own meaning and values, guiding us toward a more authentic and satisfying life.

Overcoming Cognitive Biases

Cognitive biases, such as confirmation bias and anchoring, shape our perceptions and reinforce our existing beliefs. These biases can create a distorted view of reality, limiting our ability to see things as they truly are. By embracing the absurd and adopting a flow-oriented mindset, we can become more aware of these biases and challenge our assumptions. This awareness allows us to cultivate a more open and flexible perspective, helping us to navigate the complexities of reality with greater clarity and insight.

The Role of Mindfulness

Mindfulness practices, such as meditation and deep breathing, can help us cultivate presence and awareness in the moment. By focusing on the present, we can reduce the influence of cognitive biases and connect with the deeper, more authentic layers of reality. Mindfulness encourages us to observe our thoughts and feelings without judgment, allowing us to see beyond the constructed narratives that shape our perceptions. This practice enhances our ability to embrace the absurd and go with the flow, fostering a more harmonious and enlightened way of living.

Creativity and Innovation

Embracing the absurd and practicing flowism also enhance our creativity and innovation. When we let go of rigid expectations and open ourselves to new possibilities, we can approach problems with fresh insights and creative solutions. This openness allows us to see beyond the limitations of conventional thinking and to explore new ways of understanding and interacting with the world. By integrating these philosophies, we can unlock our creative potential and discover innovative approaches to the challenges we encounter.

Building Meaningful Connections

The principles of embracing the absurd and flowism can also enhance our relationships with others. By accepting the fluid and constructed nature of reality, we can approach our interactions with greater empathy and understanding. This perspective encourages us to listen actively and to respond with flexibility and openness. By building relationships based on these principles, we can create deeper and more meaningful connections, fostering a sense of community and support.

Embracing Change and Uncertainty

Life is inherently unpredictable, and attempts to control or foresee every outcome can lead to stress and disappointment. By embracing the absurd and adopting a flow-oriented mindset, we can learn to accept and even welcome uncertainty as a natural part of the human experience. This acceptance fosters a sense of peace and reduces anxiety, allowing us to navigate life's ups and downs with greater ease and adaptability. It empowers us to face the unknown with confidence, knowing that we can create meaning and purpose even in the face of uncertainty.

Chapter 11 (The Finale) Living with the Illusion

The Practical Implications

Understanding and acknowledging the constructed nature of reality has profound practical implications for how we live and interact with the world. This recognition encourages us to cultivate open-mindedness, engage in critical thinking, and develop a willingness to question established norms and beliefs. By embracing these principles, we can foster personal growth, social progress, and a more harmonious coexistence with others. This chapter explores the various ways in which recognizing the constructed nature of reality can transform our lives and interactions.

Open-Mindedness: Expanding Horizons

Recognizing that reality is a construct encourages open-mindedness. When we understand that our perceptions and beliefs are shaped by various cultural, social, and cognitive factors, we become more willing to consider alternative viewpoints and experiences. Open-mindedness allows us to step outside our comfort zones and explore new ideas, leading to personal growth and a broader understanding of the world. It fosters a sense of curiosity and wonder, driving us to seek out diverse perspectives and experiences.

Critical Thinking: Analyzing and Questioning

Critical thinking is essential for navigating a constructed reality. By questioning our assumptions and analyzing the underlying structures of our beliefs, we can develop a more nuanced and accurate understanding of the world. Critical thinking involves examining evidence, evaluating arguments, and considering the implications of different perspectives. This process helps us to identify and challenge biases, misinformation, and flawed reasoning, enabling us to make more informed decisions and judgments.

Challenging Established Norms

Recognizing the constructed nature of reality encourages us to question established norms and beliefs. Many societal norms and conventions are based on historical, cultural, and social constructs that may no longer serve us or reflect our current understanding. By critically examining these norms, we can identify areas where change is needed and work towards creating a more equitable and just society. Challenging established norms requires courage and resilience, but it is essential for fostering social progress and innovation.

Personal Growth and Transformation

Understanding the constructed nature of reality can lead to significant personal growth and transformation. By questioning our beliefs and assumptions, we can uncover deeper truths about ourselves and our place in the world. This self-awareness allows us to align our actions and values more closely with our authentic selves, leading to a greater sense of fulfillment and purpose. It also encourages us to let go of limiting beliefs and behaviors, opening up new possibilities for personal and professional development.

Enhancing Empathy and Understanding

Recognizing that reality is constructed fosters empathy and understanding towards others.

When we appreciate that everyone's perceptions and beliefs are shaped by their unique experiences and contexts, we become more compassionate and tolerant. This understanding helps us to build stronger, more meaningful relationships and to navigate conflicts with greater sensitivity and respect. By embracing the diversity of human experience, we can create a more inclusive and supportive community.

Promoting Innovation and Creativity

Acknowledging the constructed nature of reality can also enhance innovation and creativity. When we recognize that existing structures and systems are not immutable, we are more likely to think outside the box and explore new ideas. This mindset encourages experimentation and risk-taking, driving innovation in various fields, from science and technology to art and culture. By challenging the status quo, we can discover

novel solutions to complex problems and create
a more dynamic and forward-thinking society.

Navigating Uncertainty and Change

Understanding the constructed nature of reality
helps us to navigate uncertainty and change
with greater resilience. By accepting that our
perceptions and beliefs are fluid and adaptable,
we can approach change with a sense of
curiosity and openness. This flexibility allows
us to respond more effectively to new
challenges and opportunities, reducing stress
and anxiety. It also empowers us to embrace the
unknown and to see change as a natural and
necessary part of life.

Building a More Equitable Society

Recognizing the constructed nature of reality
has important implications for social justice and
equity. Many social hierarchies and inequalities

are based on constructed beliefs and systems that perpetuate discrimination and oppression. By critically examining these constructs, we can work towards dismantling harmful structures and creating a more inclusive and equitable society. This process involves advocating for systemic change, promoting diversity and inclusion, and supporting marginalized communities.

Cultivating a Lifelong Learning Mindset

Embracing the constructed nature of reality fosters a lifelong learning mindset. When we acknowledge that our understanding of the world is always evolving, we become more committed to continuous learning and self-improvement. This mindset encourages us to seek out new knowledge and experiences, to reflect on our growth, and to remain open to change. Lifelong learning helps us to stay adaptable and relevant in a rapidly changing world, enhancing our personal and professional lives.

Finding Meaning in a Constructed World

The acknowledgment that reality is a construct does not diminish the value of human experiences and achievements. Despite the subjective nature of our perceptions and beliefs, meaning can be found in various aspects of life, from personal relationships and creative endeavors to the pursuit of knowledge. This chapter explores how recognizing the constructed nature of reality enriches our understanding of these pursuits and underscores their intrinsic value.

The Value of Human Experiences

Human experiences, such as love, joy, sorrow, and fulfillment, are deeply meaningful regardless of the constructed nature of reality.

Our interactions with others, our emotions, and our personal growth contribute to the richness of life. These experiences shape who we are and provide us with a sense of purpose and connection. By embracing the beauty and complexity of human experiences, we can find profound meaning and fulfillment in our everyday lives.

Personal Relationships: Connections Beyond Constructs

Personal relationships are fundamental to human existence and provide a source of meaning and support. Whether with family members, friends, or romantic partners, these connections transcend the constructed narratives of reality. They offer us companionship, empathy, and a sense of belonging that enriches our lives. Building and nurturing meaningful relationships allows us to share experiences, celebrate achievements, and navigate challenges together, fostering personal growth and emotional well-being.

Creative Endeavors: Expressions of Imagination

Creative endeavors, such as art, music, literature, and innovation, are expressions of human imagination and innovation. Despite being shaped by cultural, social, and historical contexts, creativity transcends the limitations of constructed reality. It allows us to explore new ideas, challenge conventions, and inspire others. Engaging in creative pursuits not only brings joy and fulfillment but also contributes to cultural enrichment and societal progress. Creativity is a testament to the human spirit's capacity to imagine and create, regardless of the constructs that surround us.

The Pursuit of Knowledge: Uncovering Truths

The pursuit of knowledge and understanding is another meaningful endeavor that transcends the constructed nature of reality. Through scientific inquiry, philosophical exploration, and academic scholarship, we seek to uncover truths about the world and ourselves. This quest for knowledge enriches our understanding of reality, challenges our assumptions, and expands the boundaries of human understanding. It fosters intellectual curiosity, critical thinking, and innovation, driving progress in various fields and contributing to human flourishing.

Finding Meaning and Purpose

Ultimately, the pursuit of meaning and purpose is a deeply personal journey that transcends the constructed nature of reality. Whether through personal relationships, creative endeavors, or the pursuit of knowledge, individuals can find fulfillment and satisfaction in contributing to something greater than themselves. These pursuits allow us to create meaningful legacies, inspire future generations, and leave a positive impact on the world.

Embracing the Paradox

Acknowledging that reality is a construct involves embracing a paradox: while our perceptions may be subjective and influenced by cultural and cognitive factors, the experiences and achievements we value are real and meaningful to us. This paradox highlights the complexity of human existence and underscores the importance of embracing ambiguity and uncertainty. By navigating this paradox, we can cultivate a deeper appreciation for the diversity of human experiences and perspectives.

Resilience and Adaptability

Recognizing the constructed nature of reality encourages resilience and adaptability in the face of change and uncertainty. It reminds us that our understanding of the world is constantly evolving, and that we have the capacity to adapt and grow. This resilience allows us to navigate life's challenges with grace and perseverance, knowing that our experiences and achievements hold value regardless of external constructs or interpretations.

The Balance Between Skepticism and Acceptance

A balanced approach to navigating constructed realities involves both questioning and understanding the constructs that shape our perceptions, while also engaging with them pragmatically. This chapter explores how embracing this balance fosters a more nuanced and reflective approach to life, enabling us to navigate complexity with wisdom and insight.

Questioning Constructs: Cultivating Critical Thinking

Questioning the constructs that shape our reality is essential for developing a deeper understanding of ourselves and the world around us. It involves critically examining beliefs, assumptions, and societal norms to uncover their underlying motivations and implications. By questioning constructs, we can challenge entrenched perspectives, identify biases, and open ourselves to alternative viewpoints. This process encourages intellectual curiosity and fosters a more flexible and adaptive mindset, allowing us to navigate uncertainty with clarity and discernment.

Understanding Constructs: Contextualizing Perspectives

Understanding the constructs that shape our reality involves recognizing their historical, cultural, and social contexts. Constructs, such as language, laws, and social norms, provide structure and coherence to our interactions and experiences. While these constructs may be subjective and contingent, they also serve practical purposes in organizing society and facilitating communication. By understanding their origins and functions, we can appreciate their role in shaping our collective reality and navigating social dynamics effectively.

Engaging Pragmatically: Applying Constructive Insights

Engaging with constructs pragmatically involves applying insights gained from questioning and understanding them in practical contexts. It requires balancing idealism with realism, recognizing that while constructs may be imperfect or arbitrary, they influence our daily lives and interactions. Pragmatic engagement entails making informed decisions, navigating social norms, and participating in constructive dialogue. By embracing pragmatism, we can leverage our understanding of constructs to effect positive change and promote meaningful outcomes in personal and professional spheres.

Embracing Nuance: A Reflective Approach

A balanced approach to constructed realities embraces nuance and complexity. It acknowledges that reality is multifaceted and dynamic, shaped by diverse perspectives and interpretations. Embracing nuance encourages empathy, humility, and a willingness to consider multiple viewpoints. It fosters dialogue and collaboration, facilitating mutual understanding and collective problem-solving. By embracing nuance, we can navigate ambiguity with resilience and grace, recognizing the richness of human experiences and the interconnectedness of our shared reality.

Cultivating Wisdom: Integrating Insights

Balancing questioning, understanding, and pragmatic engagement cultivates wisdom—a deeper insight into the nature of constructed realities and their implications. Wisdom involves synthesizing knowledge, experience, and ethical considerations to make informed decisions and navigate ethical dilemmas. It encourages ethical behavior, compassion, and a commitment to social justice. By cultivating wisdom, we can contribute to positive societal change and promote the well-being of individuals and communities.

Resilience in Complexity: Adapting to Change

Navigating constructed realities requires resilience—a capacity to adapt and thrive in the face of uncertainty and change. Resilience involves embracing challenges as opportunities for growth, learning from setbacks, and maintaining a sense of optimism and determination. It enables us to navigate transitions, overcome obstacles, and pursue meaningful goals with resilience and determination. By fostering resilience, we can embrace the evolving nature of reality and harness its potential for personal and collective transformation.

Ethical Considerations: Acting with Integrity

Ethical considerations are integral to navigating constructed realities responsibly and ethically. They involve reflecting on the impact of our actions on others and the broader community, upholding principles of fairness, justice, and respect. Ethical behavior fosters trust, collaboration, and mutual respect, contributing to a harmonious and inclusive society. By acting with integrity, we can uphold ethical standards and contribute positively to the ongoing construction of a more just and equitable reality.

Moving Forward with Awareness

Awareness of the constructed nature of reality opens pathways to greater empathy, tolerance, and flexibility in our interactions and worldview. This chapter delves into how recognizing the fluidity and subjectivity of perceptions allows us to navigate the world with deeper understanding and appreciation for the diverse realities that coexist within it.

Embracing Fluidity: Recognizing Subjective Perceptions

Understanding that reality is constructed prompts us to acknowledge the subjective nature of our perceptions. Each individual interprets and experiences the world through their unique lens shaped by personal experiences, cultural influences, and cognitive biases. By recognizing this variability, we cultivate empathy—an ability to understand and share the feelings of others—because we appreciate that everyone's reality is valid and shaped by their distinct circumstances.

Cultivating Empathy: Bridging Perspectives

Empathy bridges the gap between different realities by encouraging us to listen actively, seek understanding, and acknowledge diverse perspectives. When we recognize that our own perceptions are constructed, we become more open to learning from others and appreciating the complexity of their experiences. This empathetic approach fosters meaningful connections, reduces conflict, and promotes mutual respect in interpersonal relationships and societal interactions.

Practicing Tolerance: Embracing Diversity

Tolerance stems from a recognition of the diversity of human experiences and perspectives. Awareness of constructed realities encourages tolerance—a willingness to accept and respect differences in beliefs, values, and behaviors. By embracing diversity, we create inclusive spaces where individuals feel valued and empowered to express their authentic selves. Tolerance fosters a sense of community and collective belonging, promoting social cohesion and collaboration across cultural, ideological, and geographical boundaries.

Flexibility in Understanding: Adapting to Change

Flexibility arises from acknowledging the dynamic nature of constructed realities. As our understanding evolves and new perspectives emerge, we remain open to revising our beliefs and adapting to changing circumstances. This flexibility allows us to navigate complexity with resilience and creativity, embracing innovation and growth. By embracing uncertainty and exploring alternative viewpoints, we expand our horizons and contribute to collective learning and progress.

Navigating Complexity: Integrating Perspectives

Navigating the complexities of constructed realities requires integrating diverse perspectives and insights. Awareness of subjective perceptions encourages us to engage in constructive dialogue, collaborative problem-solving, and inclusive decision-making processes. By integrating multiple viewpoints, we enhance our understanding of complex issues, identify innovative solutions, and promote sustainable outcomes that benefit individuals and communities alike.

Enhancing Interpersonal Relationships: Building Trust and Connection

Awareness of constructed realities strengthens interpersonal relationships by fostering trust, authenticity, and mutual understanding. When we approach interactions with empathy and tolerance, we create supportive environments where communication flows openly and conflicts are resolved constructively. Building trust and connection promotes emotional well-being and enhances the quality of relationships, contributing to personal happiness and collective harmony.

Promoting Social Justice: Advocating for Equity

Empathy and tolerance are foundational to advocating for social justice and equity. Awareness of constructed realities prompts us to challenge systemic injustices, discriminatory practices, and inequalities that perpetuate marginalization and exclusion. By advocating for fairness, inclusivity, and equal opportunities, we strive to create a more just and equitable society where every individual can thrive and contribute to the common good.

Embracing Growth and Learning: Lifelong Development

Embracing the fluidity and subjectivity of perceptions encourages lifelong growth and learning. By remaining curious, open-minded, and receptive to new ideas, we expand our knowledge, deepen our understanding, and enrich our personal and professional lives. Continuous learning fosters personal growth, enhances critical thinking skills, and empowers us to navigate complexity with confidence and clarity.

In this exploration of the illusion of reality, we have journeyed through the constructs of language, mathematics, science, culture, time, space, morality, technology, and belief. By deconstructing these elements, we gain insight into the profound creativity of the human mind and the flexible nature of the world we inhabit. This awareness opens the door to new possibilities for understanding, growth, and coexistence in a world built from the fabric of our collective imagination. And if this is all constructed, and all of the things I have said here are factual. Why should you care? Acknowledge these things. Accept the fact that we don't really know anything, and this is all just speculation. So relax, watch some TV, play some video games, and listen to music. Take a break from work. Just relax. There could be a reason why we are here. Or there might not be. Who knows?